# An Introduction to Surface-Micromachining

# An Introduction to Surface-Micromachining

*by*
**Robert W. Johnstone**
**M. Parameswaran**
*Engineering Science*
*Simon Fraser University*

**Kluwer Academic Publishers**
Boston/Dordrecht/London

Distributors for North, Central and South America:
Kluwer Academic Publishers
101 Philip Drive
Assinippi Park
Norwell, Massachusetts 02061 USA
Telephone (781) 871-6600
Fax (781) 871-6528
E-Mail <kluwer@wkap.com>

Distributors for all other countries:
Kluwer Academic Publishers Group
Post Office Box 322
3300 AH Dordrecht, THE NETHERLANDS
Telephone 31 78 6576 000
Fax 31 78 6576 474
E-Mail <orderdept@wkap.nl>

Electronic Services <http://www.wkap.nl>

**Library of Congress Cataloging-in-Publication**

Title: An Introduction to surface-Micromachining
Author (s): Robert W. Johnstone & M. Parameswaran
ISBN: 1-4020-8020-4
ISBN: 1-4020-8021-2 (eBook)


*Printed on acid-free paper.*

Printed in the United States of America

*This book is dedicated to all things flat and small.*

# Contents

# Preface

In the process of teaching several courses on surface-micromachining, it has become apparent that there is a severe gap in the existing literature. In articles and in books, authors approach the subject from one of two directions: processing knowledge or device knowledge. These two approaches neglect an important bridge between the fabrication process and actual device implementation.

A large set of skills exists that is required to design surface-micromachined devices. Knowledge of a specific micromachining process is not enough. Theory specific to a particular device is not enough. Designers must connect the manufacturing process, microscale phenomena, and design data to physical form and function. Intuition, developed by experience with macroscale systems, will not help with microscale systems. These are the basic ideas that designers must understand if they are to successfully implement any device design.

We hope that this book, while it cannot be complete, will provide a unified view of surface-micromachining. Building up from the basic building block of microfabrication techniques, this book will discuss surface-micromachining design in general, and it will finish with the theory and design of concrete components.

# Acknowledgments

We would like to thank the following people for their help in the writing of this book: Sunjaya Djaja, Linnea Duke, Ian Foulds, and Bonnie Gray. We would also like to thank all the students who took the surface-micromachining course at Simon Fraser University, their successful and less than successful experiments played a role in shaping this text.

Chapter 1

# INTRODUCTION

New designers should first understand the processing steps involved in surface-micromachining, and how these basic processing steps are combined into manufacturing processes. They should then build upon this knowledge by learning the basic design techniques used in surface-micromachining. Finally, new designers should learn any specific device knowledge and theory required. Figure 1.1 outlines this hierarchy.

In addition to process knowledge and device knowledge, designers should understand the importance of micro-scale physics. While the laws of physics are unchanged, the relative importance of phenomena does change. Further, additional physical phenomena become important in micromachined devices. At the very least, a good understanding of scaling-laws is essential.

Figure 1.1 represents our current understanding of surface-micromachining. Surface-micromachining fabrication processes form the foundation simply be-

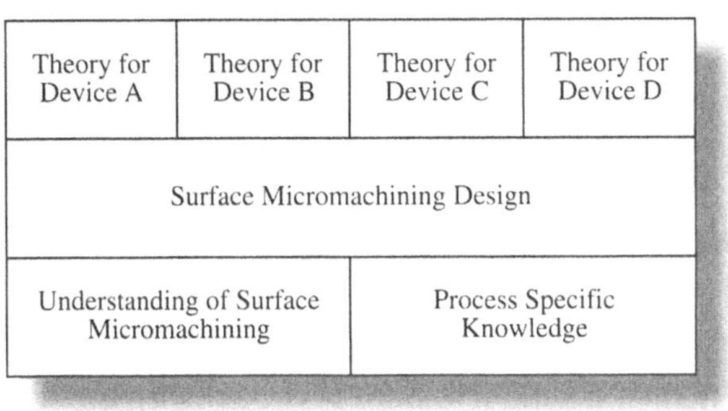


*Figure 1.1.* Knowledge layering in surface-micromachining.

cause the available processes define the design space. On top of this layer, there are the techniques of surface micromachining design. These are design concepts and approaches used in most devices. They reflect the physical reality of surface micromachined devices, and so are found, in varying degrees, in all devices. Finally, at the top, are the devices.

There is an additional layer not shown in figure 1.1, where surface-micromachined devices are collected into systems. The design and modelling of these systems is still an area of active research. Unfortunately, these topics will not be covered in this text.

With this text, we hope to provide a complete overview of surface-micromachining. In particular, the middle layer does not appear in the literature and we hope to fill in this missing middle ground. The structure of this book largely reflects figure 1.1.

## 1.1 Why miniaturize?

Historically, humans have been pushed to miniaturize many types of devices for varying reasons. The push is from pure aesthetics and fashion driving the crafting of finer and more intricate jewellery, to portability driving the transformation of grandfather clocks into wristwatches.

The miniaturization of devices, from pure electrical system to mechanical systems, is increasingly being driven by technology. Engineering criteria, such as cost, performance, and system integration provide incentives to miniaturize.

- Miniaturized components allow complete systems to be created in a single package. One important benefit of system integration is avoiding the assembly of discrete components in the final device, thus *lowering cost.*

  Additionally, batch fabrication methods can drastically reduce the cost of a system's components. For example, while manufacturers may spend quite a lot of money processing a wafer into microelectronics systems, that wafer can contain billions of transistors, making the individual transistors extremely inexpensive.

- Miniature devices can have *increased reliability*. As will be discussed in the section on scaling laws (section 4.1), mass decreases faster than structural strength. Thus miniature components can withstand higher drops and larger vibrations.

  System integration also increases reliability. The interconnections between discrete components can also fail, so monolithically manufactured systems typically exhibit greater reliability.

- Miniature devices can have *faster response times*. Smaller devices obviously have less inertia, less thermal mass, less capacitance, etc. For many types of devices, these parameters play important roles in the device's response time.

System integration, by reducing the size of interconnections, both electrical and mechanical, reduces the amount of parasitic loads on the system. Thus system integration can also improve the response times.

- Finally, smaller devices can simply *fit more places*. For example, instead of an airplane containing a single traditional gyroscope, it could contain multiple miniature gyroscopes throughout. Not only would this provide redundancy, but additional types of analyses could be performed, improving safety and performance.

## 1.2 Why surface-micromachining?

For the last two decades, micromachining has been closely associated with microelectronics fabrication processes. The use of traditional semiconductor manufacturing techniques, such as photolithography, thin-film deposition, and etching, provide a novel approach to the shaping and processing of materials to produce functional mechanical devices. Early micromachining research and development was closely tied to existing techniques provided by the already established microelectronics industry.

However, micromachining is increasingly moving to fabrication techniques developed specifically for micromachining. These include new methods of depositing and etching films, expanding the range of film thicknesses possible. Modern micromachining is also making use of materials not found in the microelectronics industry.

Several micromachining processes exist. They all have advantages and disadvantages. The three main technologies (chapter 3) are bulk micromachining, LIGA, and surface micromachining.

- *Bulk micromachining* refers to the process of selectively etching away a portion of the substrate to form free standing structures bound by a cavity (section 3.1).

- *LIGA* uses thick layers of photoresist to serve as a mould for electroplated parts (section 3.2).

- *Surface-micromachining* builds structures out of multiple thin-films, each of which are patterned using photolithography (section 3.3). A release step is used to remove some of the thin-films.

In particular, surface-micromachining is advantageous because it is so closely related to manufacturing processes used in microelectronics. Surface micromachining benefits from lots of cross-over expertise in thin-film deposition and patterning. It is also capable of creating devices with smaller features than either bulk-micromachining or LIGA.

One disadvantage of surface-micromachining is that all devices are made from thin-films. Perpendicular to the wafer surface, conventional surface micromachined devices can be too small for the desired application. It is difficult to build thicker structures, and several important technologies have been created to fill this need. Traditionally, taller structures were handled in the design by providing parts to be assembled later. The deposition of thicker films can push the engineering burden to process development, although really tall structures (mesoscopic) still require assembly.

Surface-micromachining is not the solution to all micromachining designs. Often, it is difficult to build structures large enough. In these cases, bulk micromachining or LIGA may be more suitable. However, for very small structures, surface-micromachining excels.

## 1.3 History

A history of micromachining could easily cross-over into a history of microfabrication in general. This would then include the history of microelectronics, an obviously large topic on its own. What follows is a short history that focuses on micromachining.

The idea of miniature machines is often first attributed to Richard Feynman, who gave an influential presentation on the topic in 1959 [1]. Certainly, mechanical miniaturization had preceded the talk, but Richard Feynman clearly stated a vision of aggressive miniaturization of mechanical systems, as well as some of the problems that would be encountered (section 4.1).

Richard Feynman's talk almost foreshadows the technology that has become synonymous with miniaturization. The planar batch-fabrication process, fundamental to the microelectronics industry, was developed in 1960. At this time, feature sizes were still measured in millimetres, but allowed designers to integrate many devices on a single piece of silicon.

A major early innovation was the creation of the resonant gate field-effect-transistor (FET), in the late 1960s [2, 3]. The device was similar to other field-effect-transistors, but the gate oxide was removed. Thus, the transconductivity of the transistor was controlled by the mechanical separation of the gate and the transistor body. While the resonant gate FET never saw commercial use [3], it clearly applied surface-micromachining concepts.

The development of silicon pressure sensors in the 1970s is probably the start of a sustained effort in micromachining. The devices were produced using a mixture of microelectronics techniques and bulk-etching. Unlike resonant gate FET devices, micromachined silicon pressure sensors were commercially successful [3]. This provided a commercial motivation to micromachining.

Academic interest in micromachining was sparked in 1982, when Kurt Petersen published a comprehensive paper discussing the mechanical properties of silicon, various methods of shaping silicon, and several structures [3]. This paper can be considered a watershed event. By collecting existing knowledge about silicon fabrication beyond microelectronics, a clear picture of how micromachining can be accomplished was made visible.

For surface-micromachining, one other key event must be marked. In 1993, DARPA initiated the MUMPs foundry service. This made a standard surface-micromachining process available to the research community, and so started widespread exploration of surface-micromachining. This process is still publicly available today, although through a commercial foundry.

## References

[1] R.P. Feynman. There's plenty of room at the bottom. *Journal of Microelectromechanical Systems*, 1(1):60–66, 1992.

[2] H.C. Nathanson, W.E. Newell, R.A. Wickstrom, and J.R. Davis Jr. The resonant-gate transistor. *Transactions on Electron Devices*, 14(3):117–133, 1967.

[3] K.E. Petersen. Silicon as a mechanical material. *Proceedings of the IEEE*, 70(5):420–457, 1982.

PART I

# FABRICATION

Chapter 2

# MICROFABRICATION PROCESSES

Before addressing the manufacturing technologies used in micromachining, the individual processing steps will be covered. The techniques described in this chapter provide the foundation upon which complete fabrication processes for micromachining[1] are based.

These basic steps are also important to the microelectronics industry. In fact, most of the processing steps described here were developed in the microelectronics industry and then found additional applications in micromachining. Photolithography, etching, thin-film deposition, and doping are all used extensively in the microelectronics industry. With the growth of micromachining, new techniques for shaping miniature structures have been developed, such as anisotropic etching and deep reactive ion etching (DRIE).

This chapter will address the basic individual processing steps. These steps, when properly ordered, constitute various microfabrication technologies. Further details about the microfabrication processes described in this chapter can be found in many other sources, but books by Sze [1] or Madou [2] are recommended.

## 2.1 Photolithography

Photolithography is a process of transferring a two-dimensional pattern onto a special photosensitive material called photoresist. The pattern can then be transferred into a thin-film or to the substrate. Ideally, one would transfer the pattern directly to the thin-film of interest, and this is sometimes done[2]. However, photoresist has the important property that it is sensitive to light. Photolithography is the first step in pattern transfer.

To begin, the pattern is created as opaque regions on a transparent material[3]. This then becomes the mask or reticle used in the photolithography process (figure 2.1). Although making a mask can be very time consuming and expen-

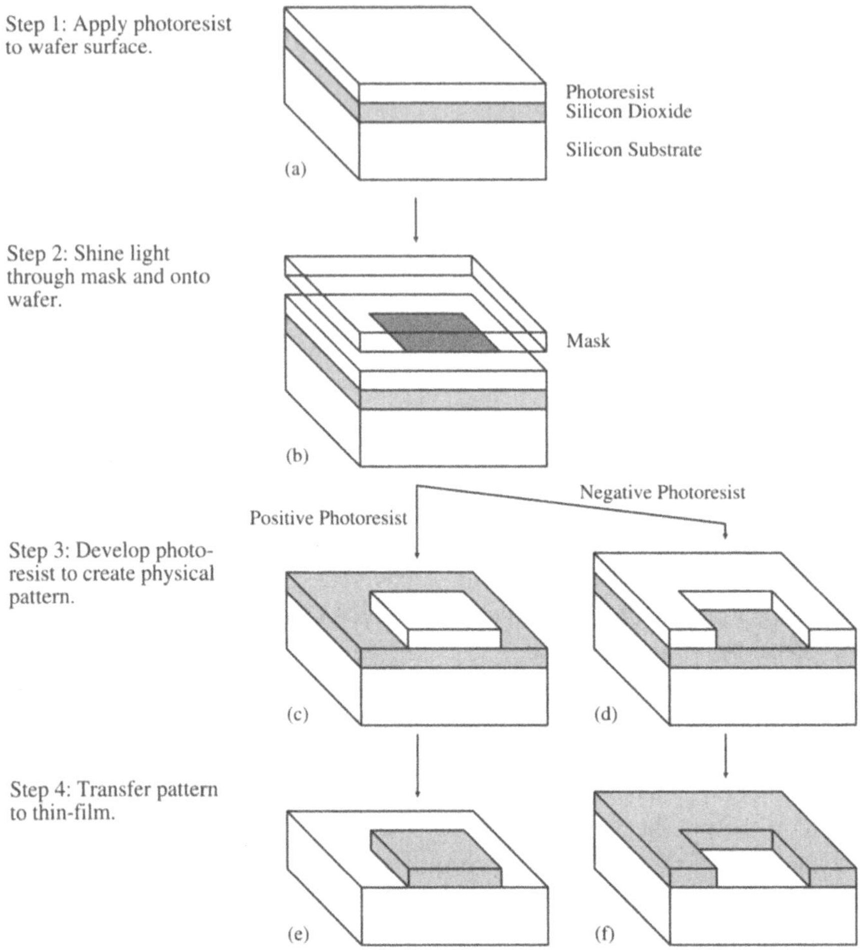


*Figure 2.1.* Illustration of a typical photolithography process. The use of silicon dioxide and silicon is only an example, and other materials are often used.

sive, masks can be used multiple times[4], and so the effort invested in fabricating the mask can be distributed across multiple wafers.

The wafer is covered with a thin layer of photoresist[5]. The photoresist is usually deposited as a few drops in the centre of the wafer. The wafer is then spun at high speed (section 2.3.1). The centrifugal force then pushes the liquid photoresist off the edge of the wafer. However, the centrifugal force is counterbalanced by the viscosity of the photoresist. By carefully controlling the speed and time that the wafer is spun, a well-controlled thickness of photoresist will be left behind.

The wafer and mask are then placed in a machine called a mask aligner. This allows the mask and the wafer to be carefully aligned, so that features from different processing steps overlap properly. Once aligned, the resist can be exposed using ultraviolet light[6] to initiate a chemical reaction. However, before reaching the photoresist, the ultraviolet light must pass through the mask. The shadows caused by the opaque regions of the mask prevent the chemical reaction in shaded regions of the photoresist. The photoresist is then developed in a chemical process similar to the photographic development.

There are two types of photoresist: positive and negative. In a positive photoresist, the exposed photoresist is washed away during the development process. The unexposed photoresist remains (figure 2.1c). In negative photoresist however, the unexposed photoresist is washed away during the development process and the exposed photoresist remains (figure 2.1d).

Once the photoresist has been developed[7], it can be used to transfer the pattern onto the underlying materials. The photoresist thus acts as a mask for subsequent processing steps.

For example, the wafer could then be used in an etching process after photolithography. The photoresist would protect portions of the wafer from the etchant. The exposed regions, however, would be attacked by the etchant and removed. In this manner, the pattern transferred to the photoresist could then be transferred to the underlying thin-film or wafer surface. After the pattern has been transferred to the wafer, the photoresist is stripped. The wafer, potentially after additional processing and cleaning, is then ready for another round of photolithography and pattern transfer.

Once a pattern has been transferred to a thin-film, it can be transferred to a lower film or the substrate. In this use, the film may be removed after pattern transfer is complete. The extra pattern transfer steps may be undesirable, but necessary due to chemical or physical constraints. In other cases, the extra pattern transfer may be desirable, such as when used for self-alignment[8].

One key advantage of photolithography is that the entire wafer is exposed simultaneously. The pattern can be transferred from the mask to the photoresist in seconds, no matter the complexity of the geometry. Similarly, any necessary chemical processing that follows photolithography also occurs simultaneously over the entire wafer surface. Thus photolithography can be used in batch fabrication processes, providing time savings.

However, with direct-write methods, the pattern is transferred to the photoresist essentially one pixel at a time. Depending on the complexity of the geometry and the writing speed, direct-write methods can be significantly slower.

### 2.1.1 Mask Aligners

Mask aligners are high precision instruments that are used to carefully align the position of a mask and a wafer. Based on the relationship between the mask

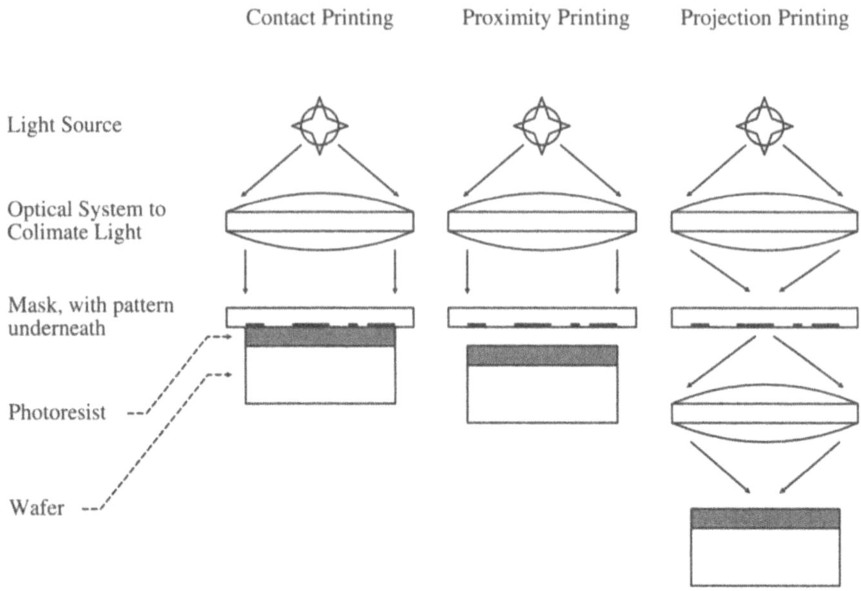


*Figure 2.2.* Types of mask aligners in common use. Parallel rays indicate collimated light, while converging rays indicate light focused to project image.

and wafer during exposure, several types of mask aligners can be identified (figure 2.2). There are three types of mask aligners:

- In *contact printing*, the mask is brought into physical contact with the wafer after alignment. The opaque regions of the mask, which are printed onto one side of the mask plate, are also in contact with wafer. During exposure, a 1:1 image is transferred to the wafer [2]. Because in contact printing there is no distance between the mask and the photoresist, there is little blurring of the shadows during exposure. The shadows' edges on the photoresist are sharp.

  However, because the mask comes into frequent contact with the wafer surface, the mask will be subject to frequent wear. Additionally, there is the risk that particulate contamination may be transferred to the mask. Both of these problems lead to errors in all subsequent pattern transfers using the same mask.

  Contact printing is common in research environments where masks are typically used a small number of times, and so wear is not a major issue.

- To avoid the problems with contact printing, *proximity printing* leaves a small gap[9] between the wafer and the mask. This gap reduces the amount of wear on the mask, allowing the mask to be used longer. However, the shadows cast by the mask blur a little in the intervening distance between the mask and the photoresist. The transferred pattern is thus also blurred, limiting resolution.

- If the mask is to be used a very large number of times, it must be completely isolated. In *projection printing*, the mask is completely removed from the wafer surface. Instead, an optical system is used to project the image printed on the mask onto the wafer surface. Additionally, the optical system usually shrinks the image[10].

  Using an optical system allows the mask to be completely removed from the wafer without necessarily sacrificing image quality, unlike proximity printing. However, the optical system significantly increases the cost of projection printing systems compared to other mask aligners.

  Often, the mask contains the image for only a small portion of the wafer. The image is thus exposed multiple times, at different locations, to fill the entire wafer.

  Projection printing is also advantageous because the image projected onto the wafer can be optically reduced. This eases dimensional and defect tolerances for the mask, usually called a reticle in projection printing. This lowers the cost of manufacturing the mask.

### 2.1.2 Resolution and Alignment

One of the key performance requirements in photolithography is resolution. Resolution is important because it partially defines the smallest piece of geometry that can be reliably transferred, and so determines the smallest dimension that can be manufactured.

Resolution of the photolithography step is not a major performance concern for surface-micromachining. The microelectronics industry has pushed the capability of photolithography techniques far into the sub-micron range. In surface-micromachining, the pattern transfer steps that follow after photolithography, when the pattern is transferred into the thin-films, limit resolution.

The resolution of photolithography system can be determined using the following equation [2]:

$$R = \frac{k\lambda}{NA} \tag{2.1}$$

In equation 2.1, $R$ is the resolution, $k$ is an dimensionless empirical constant, $\lambda$ is the wavelength of the light used for the pattern transfer, and $NA$ is the numerical aperture of the imaging system. The empirical constant, $k$,

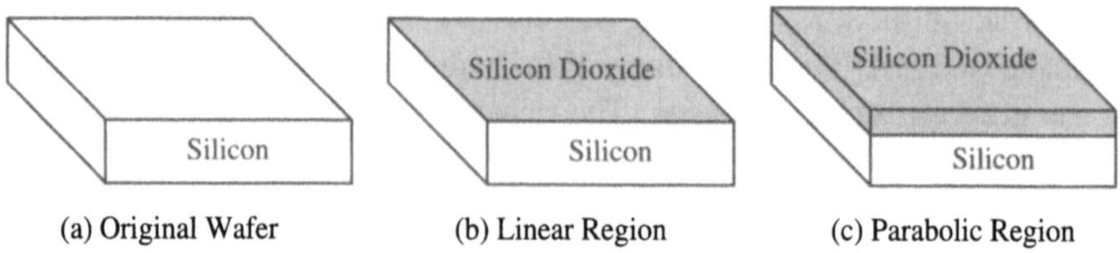


*Figure 2.3.* Thermal growth of silicon dioxide on a silicon substrate

depends on resist parameters and process conditions, and accounts for imperfections in the system. Typical values for $k$ are between 0.5 and 1.0. From optics, the value for the numerical aperture is between 0 and 1.

Alignment between different photolithography steps is also important. Again, the microelectronics industry has pushed the controls for photolithography alignment into the sub-micron range. In surface-micromachining, the pattern transfer steps that follow photolithography limit the manufacturing process.

### 2.1.3 Greyscale Photolithography

The traditional photolithography, as described above, describes binary pattern transfer. The regions of the mask are either opaque or transparent. Some micromachining techniques make use of greyscale photolithography, in which the light reaching the photoresist has multiple intensity levels.

If the light levels are properly chosen, then the chemical processes in the photoresist may undergo varying levels of completeness. The greyscale pattern on the mask can thus be transferred to the photoresist.

For example, different intensities of UV light may lead to the photoresist having varying resistance to a chemical etchant. The etching process will then remove the photoresist at different times for different regions. The underlying wafer would then also be exposed to the etchant for varying amounts of time. Thus, the etch depth could be controlled.

## 2.2 Thermal Growth

Thermal growth is a method of creating thin-films of silicon dioxide. Silicon wafers are placed in a furnace. Typically, nitrogen gas is used as an inert atmosphere while the silicon wafer reaches the process temperature. Once at temperature, an oxidizing agent, either oxygen or water vapour, is introduced. The oxidizing agent reacts with the silicon to create silicon dioxide (figure 2.3).

Thermal oxides that are very dense can be grown relatively quickly. The films contain few defects. As such, they are an important processing step in microelectronics.

The rate at which a thermal oxide grows passes through two regions: a linear rate region and a parabolic rate region. At the beginning of the oxidation process, the surface of the silicon substrate is exposed to the gases in the furnace. With the introduction of the oxidizing agent, the chemical reaction takes places initiating the growth of silicon dioxide. At the start, the rate of growth is limited by the reaction kinetics and the silicon dioxide layer increases in thickness linearly with time ($x \propto t$).

To continue the growth of silicon dioxide, the oxidizing agent must be transported to the silicon portion of the wafer. As the oxide thickens, it becomes increasingly difficult for oxygen to diffuse through the oxide layer to the silicon substrate. At some point, the rate of oxide growth becomes limited by the diffusion of oxygen through the oxide layer. For larger growth times, the thickness increases with the square root of time ($x \propto \sqrt{t}$). It therefore becomes increasingly time consuming to grow very thick films.

The main disadvantage of thermally grown oxide is that silicon from the substrate is consumed during growth. For a thermal oxide with thickness $x$, $0.44x$ of silicon is consumed [1] in the reaction. This basically limits thermal growth in surface-micromachining to the very first thin-film to be created.

Thermal growth is also inappropriate for thick films. Because of the relationship between film thickness and time, thick films can require too much time to grow thermally and so become prohibitively expensive. Alternative methods of film growth, such as chemical vapour deposition (CVD), are then preferred.

## 2.3 Deposition

In microfabrication, deposition refers to the many different processes used to add a thin-film to one or both sides of substrates.

### 2.3.1 Spin-On

The material for a thin-film is dissolved into an aqueous solution. This solution is then dispensed as a droplet on the wafer surface, near the centre. The wafer is then spun at a high speed for a fixed amount of time. During the spin, the liquid is uniformly distributed by a combination of centrifugal and viscous forces, leading to a thin aqueous film on the wafer surface. Spin-on deposition is an important step in photolithography (section 2.1).

For photoresist deposition, the following equation describes the thickness of the film after spinning [2].

$$T = \frac{KC^{\beta}\eta^{\gamma}}{\omega^{\alpha}} \tag{2.2}$$

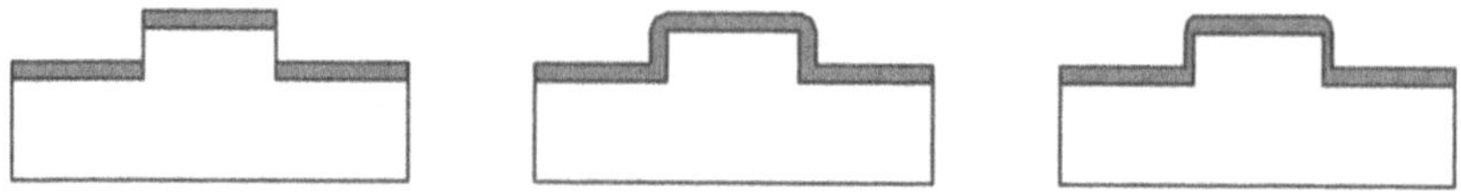

(a) Snowfall (no growth on sidewalls) (b) Conformal (equal growth on sidewalls) (c) Intermediate

*Figure 2.4.* Types of thin-film deposition based on sidewall growth. Snowfall is typical of physical deposition, while sidewall growth is typical of chemical vapour deposition.

In equation 2.2, $T$ is the film thickness, $K$ is the calibration constant, $C$ is the polymer concentration, $\eta$ is the intrinsic viscosity, and the exponential factors ($\alpha$, $\beta$, and $\gamma$) need to be determined.

Once the aqueous film has been deposited, heat is used to evaporate the solvent. In the case of photolithography, these are referred to as baking steps[11], and serve to harden the photoresist, both physically and chemically.

In the case of depositing metals or dielectrics, a further high temperature step can be used to anneal or sinter the thin-film. This is important since the spin-on films, as initially deposited, tend to be very porous.

### 2.3.2 Chemical Vapour Deposition (CVD)

Chemical vapour deposition occurs in medium to low vacuum conditions. Chemical precursors, typically gases, flow across the wafer surface[12]. The precursors chemically react to create the desired thin-film material plus other by-products. In a well designed CVD process, the thin-film materials stick to the wafer surface, where they accrete to create the thin-film while the by-products all outgas and are removed.

Often, the wafer is heated during deposition. Molecules accreting on the wafer surface, through thermal energy, then move around the wafer surface. This not only reduces porosity, but also creates more uniform coverage (figure 2.4).

The quality of the deposited films is also improved in heterogeneous reactions, where the chemical reaction takes place near the wafer surface (surface mediated). Homogenous reactions, where the chemical reaction takes place in the gas phase, create relatively poor quality films.

Several variants of CVD occur. In particular, some variants use methods to provide additional energy to the chemical reactants, thus increasing the chemical reaction rate and thus the deposition rate. Without the additional energy source, reactions must rely solely on thermal energy. For example:

- In *atmospheric pressure chemical vapour deposition* (APCVD), the reaction chamber is at (or near) atmospheric pressure.

- In *low-pressure chemical vapour deposition* (LPCVD), the reaction chamber is at a reduced pressure.

- In *plasma-enhanced chemical vapour deposition* (PECVD), the input gases are turned into a plasma by adding RF energy.

- Light- or *laser-assisted chemical vapour deposition* use photons to provide additional energy. The use of a laser also allows direct write applications.

### 2.3.3 Physical Vapour Deposition (PVD)

Physical vapour deposition occurs in very low vacuums. In PVD systems, the density of gas molecules is so low that most molecules can travel across the reaction chamber without interacting with any other molecules. This is called the free molecular regime.

One can check if the PVD chamber is in the free molecular regime by determining the Knudsen number. The Knudsen number is the ratio of the mean free path of molecules in the chamber to the characteristic size of the chamber.

$$Kn = \frac{\lambda}{L} \tag{2.3}$$

In equation 2.3, $Kn$ is the Knudsen number, $\lambda$ is the mean free path (section 4.5.2), and $L$ is the characteristic length. For a PVD system, $L$ is taken to be the size of the reaction chamber.

In a PVD system, $Kn$ should be greater than 10, indicating the system is in the free molecular regime. This means that molecules travel, on average, 10 times the size of the chamber before encountering another molecule.

- In *evaporation systems*, the source material is converted into a liquid, which then evaporates into the very low vacuum. The evaporated atoms move across the reaction chamber where the hopefully adhere to the wafer.

  For example, in thermal evaporation, the source material is placed in a crucible, which is heated. However, the crucible material may contaminate the source or, because it is also at high temperature, thermally eject atoms. To reduce contamination, electron beam evaporation can be used. In electron beam evaporation, only the centre of the target is sufficiently hot to melt. This reduces the number of non-source atoms crossing the reaction chamber, increasing the purity of the deposited film.

- In *sputtering systems*, atoms are physically dislodged from the source material. The source material is often called the target.

| | |
|---|---|
| DC Sputtering | DC voltage between target and substrate, used for conductive targets (metal films) |
| RF Sputtering | RF voltage between target and substrate, used for depositing insulators (dielectrics) |
| Magnetron Sputtering | Magnetic field confines electrons near the target, increasing the number of electrons causing ionization collisions and, thereby, deposition rates |
| Reactive Sputtering | Sputtering a target material in presence of a reactive gas, depositing a compound. |

### 2.3.4 Electrodeposition

Electrodeposition uses electrochemistry to create thin films, typically of metals. The wafer is placed in a solution containing ions of the desired material. An electric current is then passed through the solution, using the wafer as one of the electrodes. If the voltage polarity and the material of the second electrode are properly chosen, the desired ions will move towards the wafer and accrete to form a thin-film.

An interesting use of the electrodeposition is to use electrical lines that already exist on the wafer surface. One can then control where on the wafer surface electrodeposition occurs.

## 2.4 Etching

An etchant is a chemical that reacts with a solid. The products of the chemical reaction are either in either a liquid or a gaseous state, which then moves away from the solid. As the reaction continues, the solid is consumed, or etched.

The proper etchant can be a powerful tool for shaping materials [3]. The key criterion is to find an etchant that attacks the correct materials at the correct speeds. Finding an etchant that, for example, etches one material while ignoring other materials is an important part of developing a surface-micromachining process.

The rate at which a chemical etches a solid can depend on many factors. Primarily, the rate depends on the chemical composition of the material. However, the etch rate may also depend on (1) crystal orientation, (2) doping or impurity concentration, and (3) electrical potential (section 2.8).

### 2.4.1 Isotropic Etching

Most etchants are isotropic, which means that the chemical attacks all directions equally. The etch rate is independent of direction. Cavities created through isotropic etching are characterized by rounded edges (figure 2.5), since the etch rate is mostly dependent on availability of the reacting chemical.

*Figure 2.5.* An example of a cavity etched using an isotropic etchant.

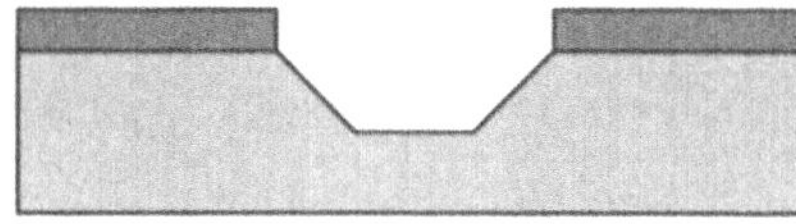

*Figure 2.6.* An example of a cavity etched using an anisotropic etchant. The substrate's orientation is (100).

Isotropic etchants are characterized by their etch rate, which will depend on the etchant's concentration and the material being etched. For example, an etchant with an etch rate of 1 $\mu$m/min will remove 1 $\mu$m each minute from a flat surface. However, at the microscale, secondary effects can be important. For example, diffusion may limit the rate at which the etchant's molecules can reach portions of the surface, thus slowing etching. This is referred to as micro-loading.

### 2.4.2 Anisotropic Etching

Anisotropic etchants have etch rates that differ depending on crystallographic direction. Crystallographic directions which etch slowly[13] are uncovered during etching as the etchant removes material in the other crystallographic directions. Materials etched anisotropically are characterized by faceted edges (figure 2.6).

In particular, silicon wafers are crystalline. Several etchants are known that are sensitive to crystal orientation. In particular, the (111) planes are more slowly etched than the (100) planes. Four of the most common anisotropic etchants for silicon are: potassium hydroxide (KOH), ethylene diamine pyrocatechol (EDP), tetramethyl ammonium hydroxide TMAH, and a mixture of hydrofluoric, nitric, and acetic acids (HNA).

In (100) oriented silicon, anisotropic etching typically leads to 'V'-grooves and pyramids, since these shapes are bounded by the (111) planes [3]. In (110) oriented silicon, the (111) planes lie perpendicular to the wafer surface [3]. This allows the etching of square structures into the substrate.

For more details, please refer to the section on bulk micromachining (section 3.1).

### 2.4.3 Reactive Ion Etching (RIE)

In reactive ion etching (RIE), the etchant is a plasma. By applying an electric field, ion motion of the etchant can be controlled. Since the ions move in a specified direction, the etching is also directed.

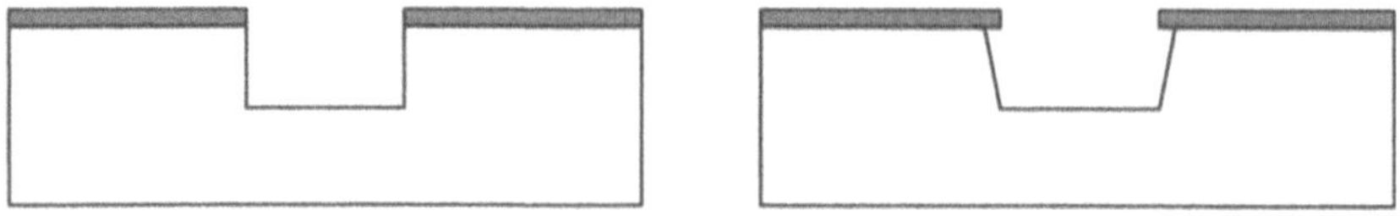

(a) Ideal sidewalls created during RIE. (b) Sidewalls showing undercut during RIE.

*Figure 2.7.* Illustration of pattern transfer using reactive ion etching. Grey regions represent photoresist.

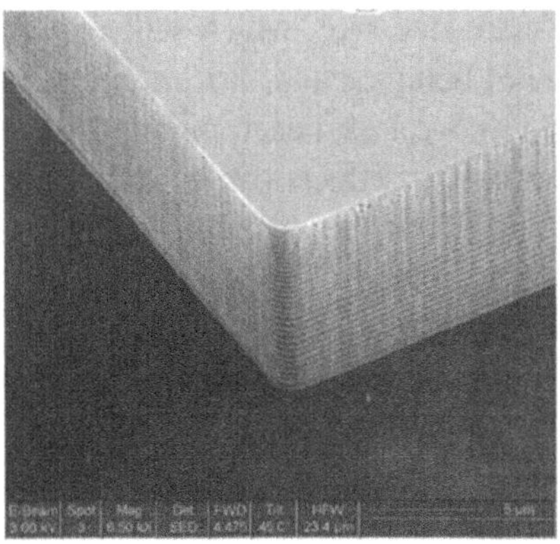

*Figure 2.8.* SEM of corner of a DRIE etched structure. Note that sidewall is textured. Fabricated using SoiMUMPs™.

This allows structures with walls perpendicular to the wafer surface to be etched. RIE is an important process in surface micromachining, but is of limited use in bulk micromachining. While the etching is predominantly vertical, lateral etching occurs at a small but measurable rate. This limits the depth that can be etched without significant undercutting.

Lateral etching occurs for a number of reasons. One reason that lateral etching occurs is that the ion motion is not perfectly controlled, so the etchant ions do have some lateral motion. The amount of lateral etch depends on the time that the sidewall was exposed, so that RIE is often characterized by walls that are sloped (figure 2.7). This limits the minimum size of structures that can be reliably patterned (section 7.2).

## 2.4.4 Deep Reactive Ion Etching

Deep reactive ion etching (DRIE) refers to a large number of processes designed to etch deep vertical structures. DRIE processes are more complicated than RIE, and include additional controls ard sub-steps that reduce the amount of sidewall etching. An examples of a DRIE process is SCREAM [4].

## 2.5 Doping

Silicon and other semiconductors are valuable as electrical materials because their electrical properties can be easily engineered. Hence, the are important in CMOS and other microelectronics fabrication processes [5, 6]. The electrical properties are controlled by introducing fixed amounts of known impurities into the material. The impurities, since they are in the silicon on purpose, are referred to as dopants. The process of adding these impurities to a semiconductor is called doping.

A pure semiconductor is also known as an intrinsic semiconductor. The charge carriers present exist solely due to thermal energy, which can move an electron from the valence band to the conduction band. This process creates two charge carriers, the electron, which is free to move around in the conduction band, and a hole in the valence band, which also acts like a charge carrier. At room temperature, thermal energy can only excite a small number of electrons to the conduction band, and so intrinsic semiconductors are poor conductors.

A doped semiconductor is also known as an extrinsic semiconductor. The dopant atoms provide additional states in which the electrons can reside. Depending on the energy level compared to the valence or conduction bands, these additional electron states can promote the creation of either free electrons or free holes, and so increase conductivity.

With sufficient doping, semiconductors can be made to act like a metal, although they will still have relatively low conductivities compared to true metals.

### 2.5.1 Diffusion

The simplest method of doping a semiconductor is through diffusion. Even in solids, the component atoms move around, and occasionally change positions. For example, thermal vibrations may lead to an interstitial atom changing locations. Substitutional atoms may also move around through interactions with crystal defects or swapping positions with other substitutional atoms.

This process is random, and atoms have an average motion of zero. However diffusion does lead to the migration of atoms away from regions with high concentration to regions with low concentrations, although the rate is typically very slow at room temperature.

The motion of dopant atoms by diffusion can be modeled using Fick's law. In one dimension, Fick's law can be stated as [1]:

$$J = -D\frac{\partial C}{\partial x} \tag{2.4}$$

In equation 2.4, $J$ is the dopant flux, $D$ is the diffusivity, $C$ is the concentration, and $x$ is the coordinate. Note that both $J$ and $C$ are functions of position and time. Equation 2.4 is often combined with the law of conservation of matter to derive Fick's second law. If impurity atoms never enter or leave the system, then there are no sources or drains. This results in the following equation [1]:

$$\frac{\partial C}{\partial t} = D\frac{\partial^2 C}{\partial x^2} \tag{2.5}$$

Equation 2.5 has been solved for a number of important boundary conditions. Readers are referred to Sze [1] or Streetman [5]. As can be seen from Fick's second law, in the absence of source or sinks the distribution of dopant atoms tends to smooth itself out over time. This process is accelerated at higher temperatures.

With sufficient time, dopant atoms can diffuse through the semiconductor. This leads to changes in impurity concentrations, and thus causes changes in the material's electrical properties. This can obviously be disadvantageous, but also provides an important method of introducing impurities into semiconductors.

In a pure semiconductor, the dopants can be introduced using diffusion. A high concentration is provided at the semiconductor surface, and given sufficient time, the dopant will diffuse into the material.

### 2.5.2 Ion Implantation

In ion implantation, the dopant atoms are ionized and placed in an electrostatic potential. The electrostatic potential accelerates the dopant ions towards the semiconductor. The ions then impact the semiconductor surface, and embed themselves. Depending on their kinetic energy, the ions will reach different depths before being stopped. After which, the atoms will then begin to diffuse. However, diffusion is usually a very slow process at room temperature.

By measuring the current between the ionization source and the wafer, the exact number of dopant atoms implanted in the wafer can be calculated. Ion implantation can therefore be very accurate [1].

During ion implantation, the dopant ions significantly damage the crystal structure. Ion implantation is therefore followed by an annealing step to repair the crystal damage and electrically activate the dopant atoms.

## 2.6 Chemical Mechanical Polishing (CMP)

When fabrication starts with the bare substrate, the surface is very smooth. As thin-films are deposited and patterned, the surface of the wafer becomes increasing corrugated. The height variations make both deposition and patterning increasingly difficult.

To overcome this, the wafer can be polished to create a new flat surface. In chemical mechanical polishing, a mixture of chemical and mechanical methods is used to polish the wafer surface. Mechanical polishing uses wear to perform gross polishing, while chemical methods can be used to soften the wafer surface. Further, some etchants can polish a surface to near atomic smoothness by preferentially attacking hills.

## 2.7 Wafer Bonding

In some microfabrication processes, structures are created on multiple wafers. To complete the system, the wafers must be joined together. The two most common methods of bonding two wafers together are anodic bonding and silicon fusing bonding.

### 2.7.1 Anodic Bonding

In anodic bonding, two substrates are clamped together. Further, the wafers are heated to between 300°Cand 400°C [7]. Each wafer is connected to an electrode and a large DC voltage (between 200 V and 1000 V) is applied. This creates a large electric field in the region between the wafers.

In this process, one of the wafers is glass that contains sodium atoms. The high temperature increases the mobility of these atoms, allowing them to migrate in the electric field. Moving the sodium atoms away from the bonding surface makes the bonding surface chemically reactive. The glass surface then chemically reacts with the other wafer, typically silicon. The two wafers are then joined due to the chemical bonds.

### 2.7.2 Fusion Bonding

In fusion bonding, two substrates are clamped together with a high contact force. If the wafer surfaces are very flat and clean, they will adhere due to interatomic forces (van der Walls, section 4.3.2). In practice, this means that the wafer surfaces must have surface roughnesses less than 10 Å, and the wafers must have a bow less than 5 $\mu$m [7]. This also means that the surfaces must be very clean before and during the bonding process.

Once the wafers have adhered to one another, the bonded wafers can be handled without separating. To complete the bonding process, the wafers are annealed, ensuring that a solid bond is formed between the two wafers. This typically increases the bond strength by over an order of magnitude [7].

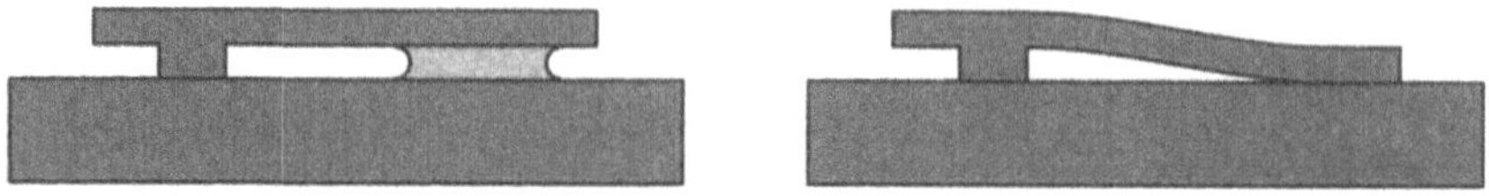

(a) Cantilever with water droplet (b) Cantilever that has adhered

*Figure 2.9.* Illustration of the drying and adhesion of a cantilever.

## 2.8 Release

The release process is specific to surface-micromachining (section 3.3). This is the step that removes all of the sacrificial materials and leaves free-standing structures.

At its simplest, the release process uses an anisotropic etchant to remove all of the sacrificial material from the wafer. Commonly, hydrofluoric acid is used since it removes silicon dioxide, while leaving silicon and most metals relatively unaffected. Once etching stops, the devices are rinsed in deionised water. The rinse removes particulate contamination and reduces the effect of solid-bridging (section 4.3.6).

Additional steps may be undertaken to change the surface chemistry of the devices. For example, these steps can ensure that the dangling bonds on silicon surfaces are hydrogen terminated or hydroxide terminated.

However, simple aqueous release processes can be very hard on micromachined systems. Many chips are designed to operate in air, so after release the chips are often removed from the liquid. As the liquid evaporates, it collects underneath devices due to capillary forces (section 4.3.6). As the liquid continues to evaporate, the liquid pulls the devices into contact with the wafer suface, leading to adhesion (figure 2.9).

To avoid the above mentioned problems, several more advanced release processes are used. These processes aim to avoid liquid evaporation.

### 2.8.1 Sublimation Drying

After rinsing in deionised water (DI) water, the devices are immersed in a liquid with a melting point near room temperature. For example, the DI water can be replaced with methanol, which can then be replaced by t-butyl alcohol. T-butyl alcohol has a melting point of 26°C. After immersion in t-butyl alcohol, the devices can be moved to refrigeration system where the t-butyl alcohol will freeze and then sublimate.

### 2.8.2 Super-critical Drying

After a traditional aqueous release, the DI water is replaced with methanol. The devices are placed in a reaction chamber where the devices are placed under pressure. The methanol is flushed with liquid carbon dioxide. The temperature is then raised to the supercritical temperature of carbon dioxide (approximately 40°C). While maintaining the temperature, the pressure in the reaction chamber is returned to atmospheric pressure. During this process, the carbon dioxide sublimates and is removed by the venting process.

The carbon dioxide leaves the wafer through sublimation, thus avoiding the liquid phase.

### 2.8.3 Vapour-phase Etching

If the etchant used to remove the sacrificial material, and the reaction by-products, are gases, then there is no liquid in the system. Vapour-phase etching thus also avoids capillary condensation. A vapour-phase etching process for silicon dioxide was presented by Lee *et al* [8].

## 2.9 Wire Bonding

In both microelectronics and surface-micromachining, the systems are embedded within or on top of a chip. Typically, the chip is encased in a ceramic or plastic package for protection. With or without the package, electrical connections to the outside world need to be established. Wire bonding is a process used to make electrical connections between the chip and the package using conducting wires.

Typically, wire bonding is performed using gold or aluminium wires. However, several different materials are available. These wires are very fine, typically only 25 to 30 $\mu$m in diameter.

The chip must contain special areas for making electrical connections called bonding pads (section 9.1), which are relatively large electrically conductive surfaces provided specifically for electrical contact.

There are four methods of wire bonding:

- In *thermomechanical wire bonding* or *thermocompression bonding*, heat and mechanical pressure are both used. The bonding equipment contains a heated chuck for the package. Once the chip has reached the proper temperature[14], the bonding equipment can be used to align the end of the wire with a bonding pad. The bonding equipment brings the tip of the wire into contact with the bonding pad and then applies pressure. The heat and pressure combine to fuse the wire with the bonding pad through metallurgical diffusion.

  The bonding machine then lifts off the bonding pad surface while feeding wire. The bonding machine can then move to another location, continuously feeding wire. The wire is then brought into contact with another bonding pad, where heat and pressure are again used to fuse the wire to the surface. The wire is then broken, leaving the bonding machine ready to create another electrical connection.

Thermomechanical wire bonding is usually performed using gold, since gold is very malleable.

- In *ultrasonic bonding*, ultrasonic energy is used to fuse the wire and bonding pad. The wire and bonding pad are pressed together and ultrasonic energy is fed to the wire. The combination of pressure and ultrasonic waves form a cold weld.

  One advantage of ultrasonic bonding is that it is a low-temperature process, and can often be performed at room temperature.

  Ultrasonic bonding is typically performed using aluminium.

- Thermomechanical and ultrasonic bonding can be combined into *thermosonic wire bonding*, where all three element are combined. Thermosonic bonding, since it uses both thermal and ultrasonic energy, can be performed at lower temperatures[15] than pure thermomechanical bonding.

  Thermosonic wire bonding is typically done with gold wire. However, copper and aluminium are also used [9].

- Finally, in *flip-chip bonding*, beads of solder are placed on the chip's bonding pads. The chip is then mounted face down in the package. The package is heated until the solder melts, forming contacts between the chip's bonding pads and the pads on the package.

  Flip-chip bonding is used for chips with a large number of pads. For wire bonding, the pads are located around the perimeter of the chip. However, in flip-chip bonding, pads can be placed throughout the chip's surface, greatly increasing the number of pads possible.

  Chips can also be wire bonded to each other using flip-chip bonding, resulting in multi-chip modules, which can be placed in a single package. This allows increased density for microelectronics. Further, multi-chip modules provide a means of integrating microelectronics and micromechanics in the same package.

  This process is also used to mount packaged chips to printed circuit boards [10].

In addition to differences in the bonding process from the mechanics of fusing the wire to the bonding pad, there are also differences in how the wire is fed during the bonding process. The two major types of bonds are wedge bonds and ball bonds.

- In *wedge bonding*, the wire is fed diagonally through the bonding equipment's tip. When the tip is pressed against the bonding pad, the tip of the wire is in between, making the first contact. The tip then moves away from the bonding site and repeats the process.

The bonding tip must move in line with the direction of the wedge. Therefore, the two bonding sites must be aligned before the bonding can begin.

- In *ball bonding*, the tip of the wire is melted and then allowed to solidify. This causes a small ball to form at the end of the wire. The ball size is 2 to 3 times the diameter of the wire. This ball is then brought into contact with a bonding pad to create the first bond. Because the tip cannot be melted again, the second bond is done using wedge bonding.

  In ball bonding, the wire is fed vertically through the tip. This allows the tip to move in any direction after the first bond has been formed. Also, the bonding tip can move faster without breaking the bonding wire in this configuration.

  However, the use of a ball bond increases the size of the bonding pad needed, and thus the bonding pads cannot be placed as closely together.

## Notes

1 This chapter should also cover most of the processing steps necessary in microelectronics processes as well.

2 For example, laser or ion milling is performed without photoresist. Laser assisted chemical vapour deposition does not need to be patterned using photoresist.

3 Typical masks substrates, often called blanks, are made from quartz or soda lime.

4 Methods of transferring a pattern directly to the photoresist, without a mask, are called direct-write methods.

5 The wafer may be covered with an adherent before the application of photoresist. The adherent is also usually deposited using spin-on techniques

6 The microelectronics industry is currently moving to X-ray light, for increased resolution.

7 Additional chemical processing, not discussed here, occurs during photolithography to harden the photoresist, both mechanically and chemically.

8 Self-alignment is an important technique in microelectronics, and ensures that the gate oxide and gate terminal are accurately aligned.

9 In the range of a few microns to a few tens of microns [2].

10 The pattern is reduced by 1:5 to 1:10 [2].

11 A soft bake is done after deposition, but before exposure. A hard bake is done after the photoresist has been developed.

12 An inert gas may also be used as a carrier.

13 The crystallographic directions which etch slowly are often called stop planes.

14 Typical temperatures for thermomechanical bonding are 300 to 400°C.

15 Thermosonic bonding occurs at temperatures around 150°C.

## References

[1] S.M. Sze (Ed.). *VLSI Technology: 2nd Edition.* McGraw-Hill, New York, 1988.

[2] M. Madou. *Fundamentals of Microfabrication.* CRC Press, Boca Raton, Florida, 1997.

[3] K.E. Petersen. Silicon as a mechanical material. *Proceedings of the IEEE*, 70(5):420–457, 1982.

[4] K.A. Shaw, Z.L. Zhang, and N.C. MacDonald. Scream i: A single mask, single-crystal silicon, reactive ion etching process for microelectromechanical structures. *Sensors and Actuators A*, 40(1):63–70, 1994.

[5] B.G. Streetman. *Solid State Electronic Devices.* Prentice Hall, Englewood Cliffs, New Jersey, 1995.

[6] S.-M. Kang and 2nd Ed. Y. Leblebici. *CMOS Digital Integrated Circuits.* WCB McGraw-Hill, Dubuque, 1999.

[7] M.A. Schmidt. Wafer-to-wafer bonding for microstructure formation. *Proceedings of the IEEE*, 86(8):1575–1585, 1998.

[8] Y.I. Lee, K.H. Park, J. Lee, C.S. Lee, H.J. Yoo, C.J. Kim, and Y.S. Yoon. Dry release for surface micromachining with hf vapor-phase etching. *Journal of Microelectromechanical Systems*, 6(3):226–233, 1997.

[9] T.C. Wei and A.R. Daud. The effects of aged cu-al intermetallics to electrical resistance in microelectronics packaging. *Microelectronics International*, 19(2):38–43, 2002.

[10] A. Sing, D.A. Horsley, M.B. Cohn, A.P. Pisano, and R.T. Howe. Batch transfer of microstructures using flip-chip solder bonding. *Journal of Microelectromechanical Systems*, 8(1):27–33, 1999.

# Chapter 3

# MICROMACHINING TECHNOLOGIES

In this day and age, everyone is familiar with microelectronics. Simply speaking, the term 'microelectronics' refers to electrical systems that have been miniaturized. However, many types of systems may be miniaturized. In particular, mechanical systems can be miniaturized as well. These miniature devices

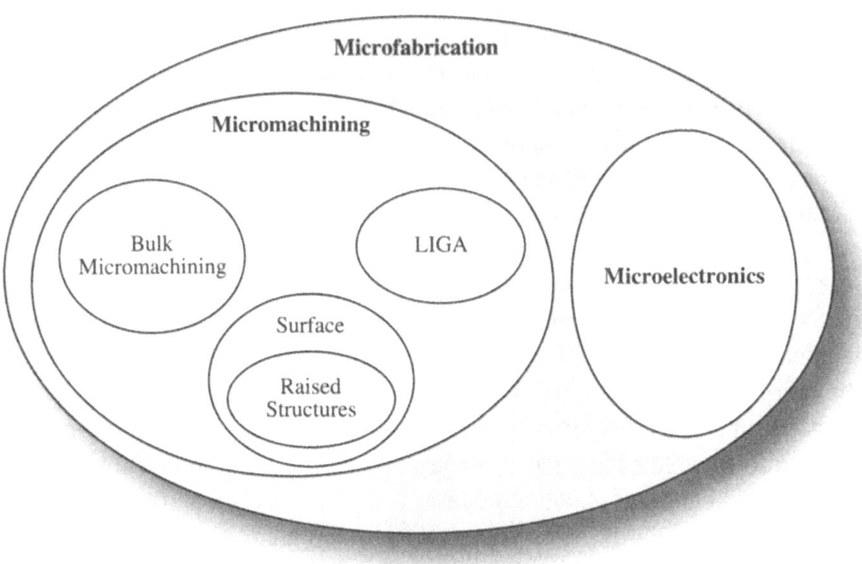


*Figure 3.1.* Relationship between various microfabrication technologies. Not all microfabrication technologies have been included.

can be called micromachines, and fabricating micromachines can be called micromachining.

The following terms are often used to describe microfabrication technology:

- Micromachining
- Microfabrication
- Microelectromechanical Systems (MEMS)
- Microsystems Technology (MST)

Most micromachining technologies, along with microelectronics, use batch fabrication methods. In a batch fabrication method, many devices are created simultaneously, saving money. For example, in microelectronics, billions of transistors are created simultaneously making the individual transistors very cheap although the processing costs for a single wafer may be significant. Similarly, micromachining hopes to make mechanical systems much cheaper by fabricating many mechanical components simultaneously.

Additionally, batch fabrication methods can not only build the components simultaneously, but also all of the interconnections. Thus, the complete system can be built avoiding the need and cost of assembling discrete components. This further lowers costs while improving reliability.

In addition to pure financial concerns, miniature devices have less inertia, less thermal mass, less capacitance, etc. For many systems, this can lead to better performance, faster response, or both. Of particular importance to mechanical systems, mass decreases faster than structural strength (section 4.1). Micromachined components therefore have the potential for increased reliability.

Despite the wide range of micromachining technologies, they share close links with the microelectronics industry. Many of the processes used in micromachining were pioneered for microelectronics. This allows micromachining fabrication to draw on established expertise and equipment. For the future, it is hoped that the close links between microelectronics and micromachining will lead to manufacturing technologies capable of creating both mechanical and electrical elements. Such a process would be ideal for sensors, actuators, and other combined types of systems. System integration, where increasingly complicated systems are built at fixed costs, is the ultimate goal.

There are several different micromachining technologies. Although the list is not exhaustive, in general, there are three main classes of micromachining technologies: bulk micromachining, LIGA, and surface-micromachining. Each of these technologies will be discussed.

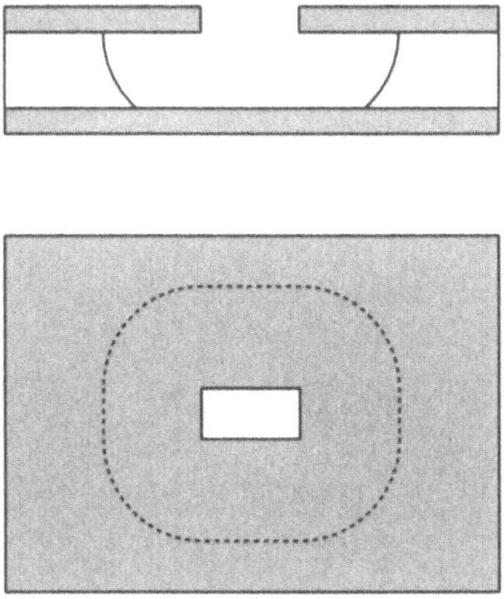

*Figure 3.2.* Example design for a membrane micromachined with an isotropic etchant.

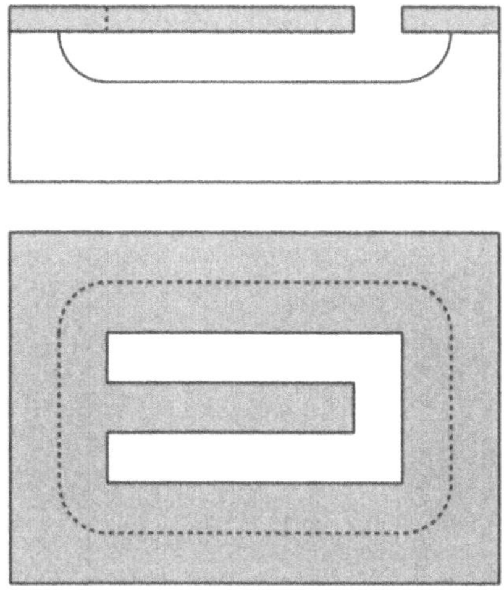

*Figure 3.3.* Example design for a cantilever micromachined with an isotropic etchant.

## 3.1 Bulk Micromachining

Bulk micromachining is a relatively crude, but valuable, method of creating silicon structures. Bulk micromachining consists of etching a substrate, typically to create cavities, pits, membranes, and cantilevers. The etching processes were discussed in the last chapter, while this section will briefly show how various structures can be fabricated.

Bulk micromachining typically starts with a silicon wafer. However, while silicon is certainly an important material, many other materials are also possible[1]. The method of etching further defines the type of bulk micromachining that occurs.

One major advantage of bulk micromachining is that it can be combined with microelectronics processes relatively easily.

### 3.1.1 Isotropic Etching

Isotropic etching is not used frequently in bulk micromachining because of the limited number of structures possible with this method. Instead, anisotropic etching is used. However, it is still possible to create membranes or cantilevers using only isotropic etching.

Referring to figures 3.2 and 3.3, the etchant attacks the substrate in a wide area around the device. This not only weakens the structure, but takes up unnecessary space.

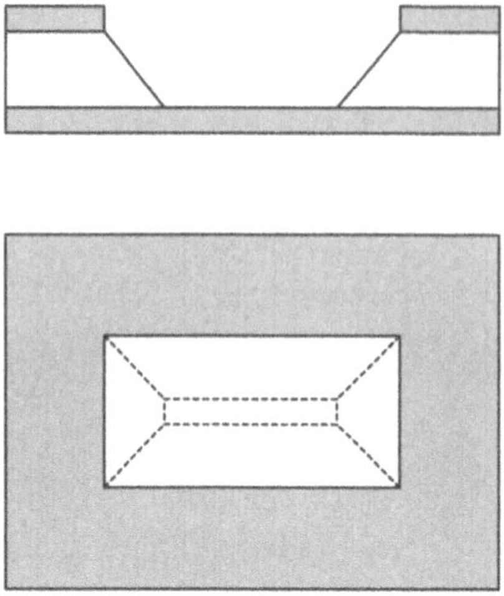

*Figure 3.4.* Example design for a membrane micromachined with an anisotropic etchant. Substrate has (100) orientation.

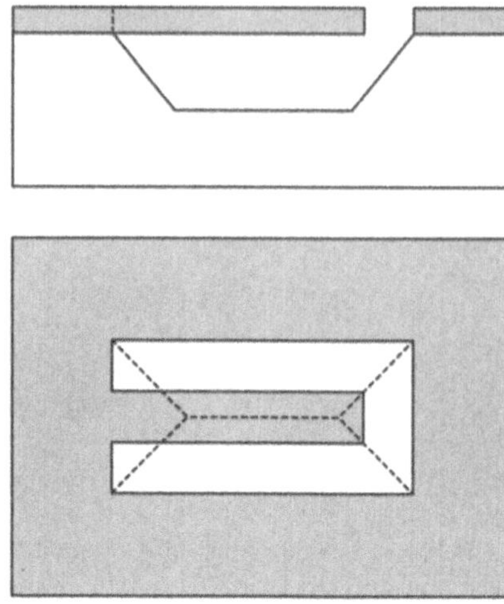

*Figure 3.5.* Example design for a cantilever micromachined with an anisotropic etchant. Substrate has (100) orientation.

### 3.1.2 Anisotropic Etching

Using an anisotropic etchant for bulk micromachining allows crystallographic planes to be used as etch stops to control the shape of the etched regions. To make proper use of the crystallographic planes, the crystal orientation has to be fixed before etching begins. Fortunately, silicon wafers come in three common orientations[2]: (100) oriented silicon, (110) oriented silicon, and (111) oriented silicon. For all three orientations, the listed crystallographic direction is perpendicular to the substrate surface.

In (100) oriented silicon, anisotropic etching typically leads to 'V'-shaped grooves and pyramids, since these shapes are bounded by the (111) planes [1, 2]. The (111) planes are at an angle of 54.7° to the surface. This leads to profiles as shown in figures 3.4 and 3.5.

Notice that in figure 3.5, the etchant undercuts the cantilever. Convex corners will act as an etch stop, but not concave corners [1]. Just as important, the structure should be aligned with the underlying crystal structure. If the mask is not properly aligned with the crystal structure, additional mask undercutting will occur [1, 2].

In (110) oriented silicon, the (111) planes lie perpendicular to the wafer surface [1]. This allows the etching of square structures into the substrate. However, the anisotropic etchant will not undercut a thin-film on the top surface, so it cannot be used to release cantilevers, unless the structure is etched from the back side of the wafer. Note that structures must still be aligned with the crystal orientation, or additional mask undercutting will occur.

In particular, there are three important anisotropic etchants for silicon [1]: ethylene diamine, pyrocatechol, and water (EDP); KOH and water; HF, $HNO_3$, and acetic acid (HNA). There are other anisotropic etchants, and etchant chemistries can become quite complex. A good introduction to bulk-micromachining is still Petersen's 1982 paper [1].

### 3.1.3 DRIE

In reactive ion etching (RIE), the etchant is a plasma. By applying an electric field, ion motion can be controlled. This allows structures with walls perpendicular to the wafer surface to be etched. RIE is an important process in surface micromachining, but is of limited use in bulk micromachining. While the etching is predominantly vertical, the sidewalls are also etched at a small but measurable rate. This limits the depth that can be etched without significant etching of the sidewalls.

Deep reactive ion etching (DRIE) refers to a large number of processes designed to etch deep vertical structures. DRIE processes are more complicated than RIE, but include additional controls and sub-steps that reduce the amount of sidewall etching. An examples DRIE process is SCREAM [3].

Using DRIE, structures similar to that produced by bulk-micromachining of (110) oriented wafers is possible. However, those structures need not be aligned with the crystal orientation. This means that, for example, trenches no longer need to be aligned with a crystallographic direction, but also that the profile can be varied.

### 3.1.4 Wafer Bonding

In addition to etching, bulk micromachining often makes use of wafer bonding. Wafer bonding refers to a process where two wafers are permanently joined. By etching multiple wafers individually, and then bonding them together, more complicated devices can be fabricated.

The two most common methods of wafer bonding are anodic bonding and silicon fusion bonding. Please refer to section 2.7 for details.

### 3.1.5 Etch Stops

In addition to crystallographic direction, material type is another method of controlling etching (figure 3.6). The etch rate is heavily dependent on material. This is key to surface micromachining, as etchants selective to the sacrificial material must be found. When used to control the extent of etching in bulk micromachining, material changes are often referred to as etch stops.

(a) p-type silicon (epi or implanted) over n-type silicon

(b) silicon dioxide over silicon

*Figure 3.6.* Bulk micromachined membrane using etch stops.

- *Silicon-on-insulator (SOI) Wafers:* SOI wafers are made up of at least two different materials. Typically, the manufacturer starts with a substrate and epitaxially grows a few layers on one side of the wafer. This leads to substrates that comprise mostly, for example, silicon, with a few thin-films on top. However, the quality and dimensional control of the thin-films is very high. This means that structures, etched using the thickness of the thin-films as a control can be also be fabricated with high dimensional control.

- *Doping:* In addition to pure material changes, such as the difference between silicon and silicon dioxide, some etchants are sensitive to changes in doping. A good example of this is Boron doped silicon, which at high concentrations ($> 5 \times 10^{18}\mathrm{cm}^{-3}$) stops several etchants.

- *Electrochemical:* Applying a voltage across a pn-junction will prevent some etchants from etching the n-type material.

The above techniques can be used to create etch stops parallel to the wafer surface. This makes them useful manufacturing membranes, as illustrated in figure 3.6,

## 3.2 LIGA

LIGA is a german acronym for lithography, electroforming, and moulding [4]. One of the defining features of LIGA is the use of X-ray lithography. X-ray light sources are more expensive than conventional ultraviolet light sources used in photolithography, but X-rays have a much shorter wavelength and so have higher resolution (section 2.1.2).

The first step is to deposit a thick layer of photoresist[3] [4]. The photoresist is exposed to x-rays, typically from a synchrotron. In addition to the exotic light source, special masks are also required [4]. In the next step, the developed photoresist is used as a mould for electroforming. The components are created by electrodeposition, with the material filling in the gaps in the photoresist mould. The resist is then stripped, leaving behind the metal structure, which can be used either directly or as a mould for further manufacturing processing.

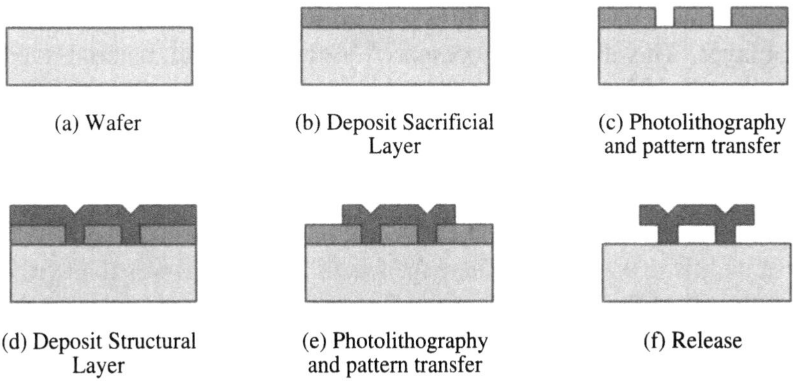


*Figure 3.7.* Example of a minimal surface-micromachining process.

## 3.3 Surface-Micromachining

Surface-micromachining involves the deposition and patterning of thin-films. Both deposition and patterning are heavily used in the microelectronics industry, providing a large base of expertise. In surface-micromachining, these techniques are adapted to build mechanical, instead of electrical, devices. Additionally, some processes use chemical-mechanical-polishing (CMP).

In surface-micromachining, alternating thin-films of two different materials are deposited. After each film is deposited, it is patterned using photolithography (section 2.1). The structural material[4] is used in structures that will remain in the final device. The sacrificial material is used to create spacers and will be removed from the final device.

Surface-micromachined devices are often referred to as planar or 2.5D devices. As you will see, most of the fabrication steps used in surface-micromachining can only act on the top surface of the wafer.

The simplest surface-micromachined device is a bridge (figure 3.7). Fabrication begins with a substrate, which provides the foundation for the fabrication process. A typical substrate is a circular piece of single-crystalline silicon with a diameter of between 3 and 8 inches. However, while silicon is popular, almost any material can be used as a substrate[5].

If the substrate is conductive, a thin-film of some insulating material is first deposited. Sticking with silicon as our example, silicon nitride is a common choice for the insulating material. While this layer is deposited, it is not usually patterned and may be thought of as part of the substrate. The insulating material allows for electrically isolated structures to be created. This layer is referred to as the isolation layer.

Surface-micromachining really begins with the deposition of the first sacrificial layer. This thin-film is composed of the sacrificial material, typically silicon dioxide. The sacrificial material is used to fill spaces that will later be empty. The sacrificial layer is then patterned using photolithography.

A thin-film of the structural material is then deposited and patterned. The structural material will remain in the final device, and is often polycrystalline silicon (polysilicon). It is important to note how the structural material, when deposited, fills in the regions where the sacrificial material was removed.

Finally, all of the sacrificial material is removed. In our example, this involves placing the silicon wafer into a bath of hydrofluoric acid, which will anisotropically etch the silicon dioxide. Hydrofluoric acid does not remove[6] the polysilicon. This is called the release process (section 2.8).

The above description is a bare minimum surface-micromachining process. Several other steps that may be added are:

- Often, a layer of the structural material is deposited before the first sacrificial layer. Since this layer is always fixed to the substrate, structures built from this layer are never free to move. However, this layer can be used to carry electrical signals, and most importantly, electrically shield devices (section 8.1). This layer is often referred to as the ground layer, or the zero layer.

- Additional layers of the sacrificial and structural material may be added. Adding more layers means more complicated devices can be built.

- After the last structural layer has been deposited, a thin-film of a metal may be deposited and patterned. This allows structures with very low resistivity to be created. The metal may or may not be a structural layer, depending on its structural integrity.

  For example, a metal layer can be deposited on top of the last structural layer in PolyMUMPs™. However, the metal can not be deposited without the underlying structural layer since it is not strong enough to exist without the support.

One set of important devices that can be built with surface-micromachining are hinges [5]. Hinges are particularly important because they allow the otherwise relatively flat surface-micromachined devices to reach the third dimension. These structures will be discussed further in chapter 13.

### 3.3.1 Sacrificial and Structural Materials

The materials used in a particular surface-micromachining process can be divided into two groups: structural materials and sacrificial materials. The structural materials are relatively untouched by the etching during the release

| Process | Structural | Sacrificial |
|---|---|---|
| MEMGen (a) | Nickel | Copper |
| MEMGen (b) | Copper | Nickel |
| PolyMUMPs™ | Polycrystalline silicon and gold | Silicon dioxide |
| MetalMUMPs™ | Polycrystalline silicon, copper, nickel and gold | Silicon dioxide |
| SUMMiT V™ | Polycrystalline silicon | Silicon dioxide |

*Table 3.1.* Materials in some surface-micromachining processes.

process and so remain in the final device. The sacrificial materials are removed during the release process and so are not present in the final device. The sacrificial materials are used to define spaces, or voids, in the structures.

It is important to note that the role a material plays in a particular surface-micromachining process depends strongly on that process. For example, surface micromachining based on copper and nickel is possible. Etchants can be found that will selectively remove either the copper or nickel. Copper and nickel can therefore swap roles of being the structural and sacrificial materials.

Since, in surface-micromachining, materials are deposited as thin-films, it is common to refer to the different films, or layers, as being structural or mechanical. Thus the various layers assume different roles.

- The *structural materials* should have good mechanical properties. For most devices, the structural material should also have good electrical properties.

- The *sacrificial materials* should be stable during deposition and throughout the fabrication process. However, it should etch quickly during the release step.

Ideally, all materials used in a surface-micromachining process are easily patterned. Handling the different materials will be much easier if they can be deposited and patterned using standard microelectronics processes.

Finally, the etchant should have a high selectivity for the sacrificial materials during release.

Table 3.1 lists some surface-micromachining processes and the materials they use. Historically, surface-micromachining processes have depended on materials readily available from microelectronics, and so processes based on silicon/silicon dioxide, silicon/silicon nitride, and aluminium/silicon nitride are common [6].

## 3.3.2 Layer Count

As more layers are added to a surface-micromachining process, additional types of devices become possible. For example, the process outlined in figure

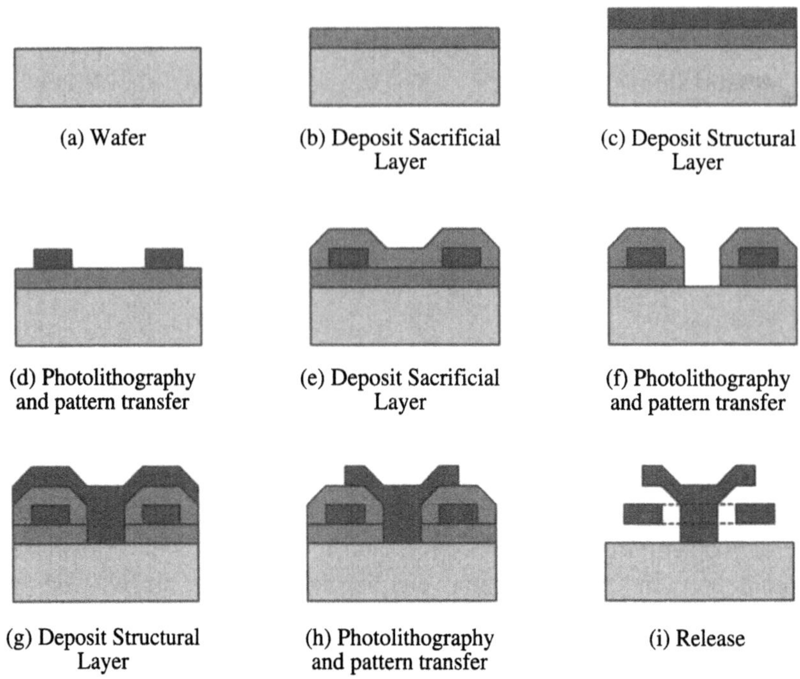


*Figure 3.8.* Example of a minimal surface-micromachining process.

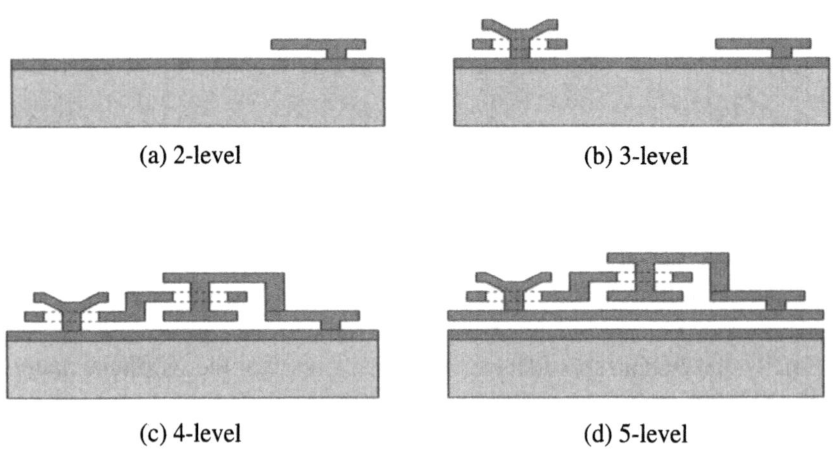


*Figure 3.9.* Example of a minimal surface-micromachining process.

3.7 is capable of creating bridges and cantilevers. To create other types of structures, additional levels are necessary.

To create wheels, gears, and other rotating parts, another structural layer is required to create a flange overtop the wheel. A sample process is illustrated in figure 3.8.

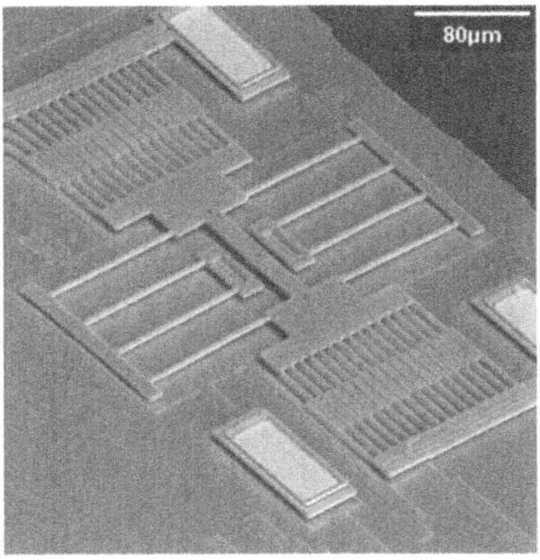


*Figure 3.10.* SEM of electrostatic comb-drive resonator, which can be built in two-layer processes. Fabricated using PolyMUMPs™.

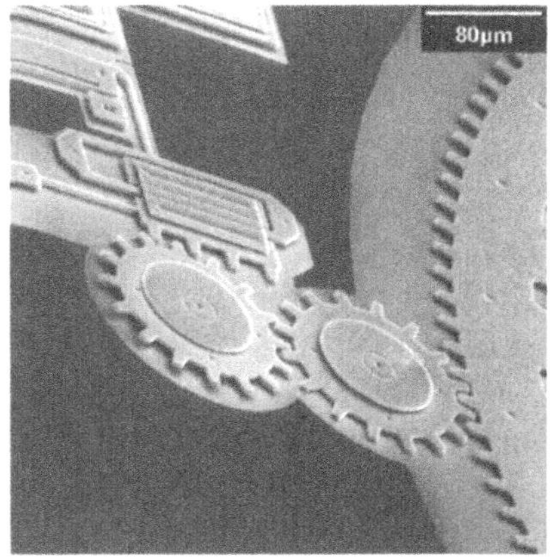


*Figure 3.11.* SEM of a gear train, which can be built in three-layer processes. Fabricated using PolyMUMPs™.

As more layers are added, more mechanical structures can be created (figures 3.9). More complicated devices, such as pin-joints and moving platform, require additional layers. For example, both PolyMUMPs™ and SUMMiT™ are surface-micromachining processes that use polycrystalline silicon and silicon dioxide. However, the PolyMUMPs™ process is a three-layer process, while SUMMiT™ has up to five structural layers. SUMMiT™ is therefore capable of creating several structures not available in PolyMUMPs™.

Additional layers require some type of planarization (section 2.6). For example, the wafer is planarized before depositing the fourth structural layer[7] in SUMMiT™.

## 3.3.3 Isolation Layer

Before the deposition of any structural or sacrificial layers, an isolation layer may be deposited. The isolation layer will be electrically isolating.

Isolation layers are necessary when the substrate is conducting. Without an isolation layer, all devices deposited on the wafer will be electrically shorted together through the substrate. Thus, an isolation layer is necessary for silicon substrates but not glass substrates.

## 3.3.4 Microelectronics

Sensors and actuators usually form low-level components of a larger system. Integrating these components with electronics is therefore an important task [7].

*Figure 3.12.* SEM of a gear containing a pin-joint, which can be built in four-layer processes. Fabricated using SUMMiT$^{TM}$. Courtesy Sandia National Laboratories, SUMMiT$^{TM}$ Technologies, www.mems.sandia.gov.

*Figure 3.13.* SEM of multilevel gears, which can be built in five-layer processes. Fabricated using SUMMiT$^{TM}$. Courtesy Sandia National Laboratories, SUMMiT$^{TM}$ Technologies, www.mems.sandia.gov.

Because of their similarities, it is possible to develop a manufacturing process which includes both microelectronics and micromechanics processes [8]. This allows extremely tight integration of the sensors, actuators, and electronics. Unfortunately, a monolithic process that includes both micromechanics and microelectronics requires many processing steps, and so these combined processes have relatively low yields[8] compared to separate microelectronics or micromachining processes.

Fortunately, there exists a range of options for systems that combine micromechanics and microelectronics. These options vary from completely separate packages, to multiple chips in the same package, to the monolithic processes just discussed. Increasing integration typically means increased performance, since parasitics and noise associated with the interconnects is reduced. However, increased integration also typically means increased cost due to increased manufacturing complexity[9].

## Notes

1 Glass has become a popular substrate material.

2 In principle, silicon substrates could be fabricated to provide any desired orientation.

3 In LIGA, the photoresist clearly must be x-ray sensitive, such as polymethylmethacrylate (PMMA).

4 In general, there may be multiple structural materials.
5 Also, substrates made from multiple materials may be used.
6 Hydrofluoric acid does etch polysilicon, but at a much lower rate than silicon dioxide.
7 The wafer is also planarized before the fifth structural layer is deposited, if the fifth structural layer is present.
8 Each process step has a yield, which is between 0 and 100%. The process yield is the product of the yields for all of the steps.
9 Tighter integration also means tighter coupling of design and manufacturing constraints.

## References

[1] K.E. Petersen. Silicon as a mechanical material. *Proceedings of the IEEE*, 70(5):420–457, 1982.

[2] Y. Backlund and L. Rosengren. New shapes in (100) si using koh and edp etches. *Journal of Micromechanics and Microengineering*, 2(2):75–79, 1992.

[3] K.A. Shaw, Z.L. Zhang, and N.C. MacDonald. Scream i: A single mask, single-crystal silicon, reactive ion etching process for microelectromechanical structures. *Sensors and Actuators A*, 40(1):63–70, 1994.

[4] A. Rogner, J. Eicher, D. Munchmeyer, R.-P. Peters, and J. Mohr. The liga technique-what are the new opportunities. *Journal of Micromechanics and Microengineering*, 2(3):133–140, 1992.

[5] K.S.J. Pister, M.W. Judy, S.R. Burgeett, and R.S. Fearing. Microfabricated hinges. *saa*, 33(3):249–256, 1992.

[6] C. Linder, L. Paratte, M.-A. Gretillat, V.P. Jaecklin, and N.F. de Rooij. Surface micromachining. *Journal of Micromechanics and Microengineering*, 2(3):122–132, 1992.

[7] A. Sing, D.A. Horsley, M.B. Cohn, A.P. Pisano, and R.T. Howe. Batch transfer of microstructures using flip-chip solder bonding. *Journal of Microelectromechanical Systems*, 8(1):27–33, 1999.

[8] J.M. Bustillo, G.K. Fedder, C.T.C. Nguyen, and R.T. Howe. Process technology for the modular integration of cmos and polysilicon microstructures. *Microsystems Technology*, 1(1):30–41, 1994.

PART II

# DESIGN

# Chapter 4

# MICROSCALE PHYSICS

The relative magnitude of different forces changes with the characteristic size of a system. At very large scales, physical problems are handled using relativity. At very small scales, physical problems are handled using quantum mechanics. Even a small change in scale, from ordinary macroscopic object to microscopic objects, can have a great effect. While the physical forces are the same, the relative importance changes considerably. Intuition based on macro-scale systems becomes unreliable when applied to micro-scale systems.

Surface-micromachining differs from conventional machining not only in the fabrication processes used, but also in fundamental operational principles. For example, the increased importance of friction and wear in micromachined design leads to very different devices. Understanding why surface-micromachined devices do not simply copy their larger brethren requires understanding micro-scale physics.

This chapter begins with an introduction to scaling laws, and tries to show how the system behaviour changes at the microscale. This chapter also covers a number of phenomena important at the microscale.

## 4.1 Scaling Laws

Scaling laws are rules used to predict how a system will behave as it changes size. In particular, we are interested in how systems behave as they are shrunk. As a general rule, forces that depend on volume, such as gravity and inertia, decrease faster than forces that depend on area, such as friction and adhesion. However, if you can write an expression for some parameter of interest, you can easily work out how it will scale.

Scaling laws can be used to make design decisions [1, 2]. For example, understanding how various actuation principles scale (electrostatic, magnetic, electro-thermal, etc.) can be used to select the most appropriate method. For example, contrary to macroscopic devices, electrostatic forces are usually preferred over magnetic forces in micromachines.

An interesting example of scaling laws is ants. More specifically, the commonly stated fact that an ant can carry 100 times its body weight. After a little investigation, this fact is not as amazing as it first appears. The creature's muscles can exert a force proportional to their cross-sectional area. However, a creature's mass is proportional to its volume. An equation relating these quantities can be written as:

$$R = \frac{\text{mass that can be lifted}}{\text{mass of creature}} \tag{4.1a}$$

$$= \frac{c_1 A}{c_2 V} \tag{4.1b}$$

To understand how $R$ changes as our creature changes size, assume that all of its dimensions are scaled equally by the parameter $s$. The scaling law for relative lifting capacity, $R$, is therefore given by the following:

$$R' = \frac{c_1(s^2 A)}{c_2(s^3 V)} \tag{4.2a}$$

$$= s^{-1}\frac{c_1 A}{c_2 V} \tag{4.2b}$$

$$= s^{-1}R \tag{4.2c}$$

From equation 4.2c, we can see that relative lifting strength is inversely proportional to scale. An ant may be able to lift 100 times its own body weight, but it is also approximately 100 times smaller than a human.

As will be seen, forces typically scale as $F \propto s^1$ to $F \propto s^4$. Forces that scale as $s^1$ or $s^2$ are valuable since they show improved response times.

### 4.1.1 Gravitational Forces

The force of gravity between two bodies is a well known physical law. All one needs to know is the mass of the two bodies and the distance between their centres of mass. However, the standard expression (equation 4.3a) does not show how the force depends on the system's dimensions. When working out the scaling law for mass, one must remember that mass is related to volume.

$$F_g = \frac{G m_1 m_2}{r^2} \tag{4.3a}$$

$$= \frac{G\,(\rho_1 V_1)\,(\rho_2 V_2)}{r^2} \tag{4.3b}$$

*Figure 4.1.* Singly-supported horizontal beam.

In equation 4.3a, $F_g$ is the force due to gravity, $G$ is the Gravitational constant, $m_1$ and $m_2$ is the mass of the two objects, and $r$ is the distance separating the centres of mass. To consider scaling, the two masses are expressed in terms of their average density, $\rho_1$ and $\rho_2$, and their volumes, $V_1$ and $V_2$.

When miniaturizing a system, the first mass will most likely be the earth. Hopefully, no one is planning to miniaturize this component of the system, so only $V_2$ is shrunk. Under this assumption, the scaling law for gravitational force is:

$$\begin{aligned} F_g' &= \frac{G\rho_1 V_1 \rho_2 \left(s^3 V\right)}{r^2} \\ &= s^3 F_g \end{aligned} \tag{4.4}$$

The scaling of gravitational force is not unexpected. As previously stated, forces that are proportional to volume typically scale as $s^3$.

The force of gravity is not necessarily of primary interest. Usually, the interesting property is how the system responds to gravity. For example, if the system were a singly-supported horizontal beam (figure 4.1), then the beam's deflection due to gravity might be important. If the deflection is relatively small, then the maximum deflection can be determined using the following equation:

$$\delta_{\max} = -\frac{gwt\rho L^4}{8EI} \tag{4.5a}$$

$$= -\frac{3g\rho L^4}{2t^2} \tag{4.5b}$$

In equation 4.5a, $\delta_{\max}$ is the maximum deflection, $g$ is the acceleration due to gravity, $w$ is the width of the beam, $t$ is the thickness of the beam, $\rho$ is the density of the beam, $L$ is the beam length, $E$ is Young's modulus, and $I$ is the bending moment of inertia. Substituting for $I$, one reaches equation 4.5b.

Assuming all dimensions are equally scaled, one can work out the following scaling law for the peak deflection:

$$\delta'_{\max} = -\frac{3g\rho(sL)^4}{2(st)^2}$$

$$
\begin{aligned}
&= -s^2 \frac{3g\rho L^4}{2t^2} \\
&= s^2 \delta_{\text{max}} \qquad (4.6)
\end{aligned}
$$

From equation 4.6, one can see that the maximum deflection shrinks faster than the device itself. Designers would more likely be interested in the relative peak deflection, $\delta_{\text{max}}/L$, which scales as $s$.

Gravity is only one type of acceleration that a device may undergo. The results of this section are easily extended to the case of general acceleration. An important consequence of equation 4.6 is that micromechanical devices can withstand larger accelerations without breaking than macroscopic mechanical systems [3].

### 4.1.2 Electrostatic Forces

Electrostatic forces are particularly important for miniature devices because of their favourable scaling. However, the scaling law changes depending on whether one assumes a constant voltage or a constant electric field.

At large scales, the electrostatic force is limited by dielectric breakdown, where the dielectric can be either air or an insulating material. Using two parallel plates as our example, the electrostatic force can be expressed as:

$$
F_e = \frac{\epsilon A V^2}{2d^2} \qquad (4.7)
$$

In equation 4.7, $F_e$ is the electrostatic force, $\epsilon$ is the permittivity of the intervening material, $A$ is the area of the plates, $V$ is the electric field, and $d$ is the distance separating the plates. Since we are interested in keeping the electric field constant, the ratio $V/d$ must remain constant. This leads to the following scaling law:

$$
\begin{aligned}
F'_e &= \frac{\epsilon \left(s^2 A\right)}{2} \left(\frac{V}{d}\right)^2 \\
&= s^2 F_e \qquad (4.8)
\end{aligned}
$$

At small scale, dielectric breakdown becomes less important (section 4.5.3). Instead, it is more convenient to operate under constant voltage. An example is a microdevice operating under external voltage control. In this case, the voltage remains fixed and $d$ must be scaled along with $A$.

$$
\begin{aligned}
F'_e &= \frac{\epsilon (s^2 A) V^2}{2(sd)^2} \\
&= s^0 F_e \qquad (4.9b)
\end{aligned}
$$

When the voltage is held constant, the electrostatic force is independent of scaling. This is extremely good and is one of the reasons why electrostatic forces are so important in microdevices.

### 4.1.3 Magnetic Forces

Permanent magnets maintain their strength, as measured by the magnetic flux density, when miniaturized [4]. The field strength therefore is scale invariant ($s^0$). However, the total flux will also depend on the cross-sectional area ($s^2$).

There are three cases to be examined for electromagnetic scaling: constant current density, constant heat flux through the conductor surface, and constant temperature increase in the conductor [4]. The derivation of these scaling laws will not be included here, but can be found in [4].

When the current density is kept constant[1], the force generated between two coils scales as $F_m \propto s^4$.

If heating is the limiting factor, then constant heat flux might be more appropriate. In this case, the current density scales as $J \propto s^{-0.5}$ and the magnetic force between two coils scales as $F_m \propto s^3$.

The final case is constant temperature rise, which might be appropriate if the melting point of some component of the system is a limiting factor. In this case, the current scales as $J \propto s^{-1}$, and magnetic forces scales as $F_m \propto s^2$.

Constant temperature increase might be used when the conductors are insulated, and the insulator's maximum temperature[2] is a limiting factor. In this case, the force current density scale as $J \propto s^{-1}$. The force between interacting electromagnets then scale as $F_m \propto s^2$.

### 4.1.4 Response Time

In the preceding sections, the scaling of a number of different forces have been explained. However, as already mentioned, the forces are not often of direct importance. One important parameter is the response time. For mechanical systems, the response time can be determined from the transit time.

$$t = \sqrt{\frac{2x}{a}} \tag{4.10a}$$

$$t = \sqrt{\frac{2xm}{F}} \tag{4.10b}$$

In equation 4.10a, $t$ is transit time of the system, $x$ is the distance that needs to be travelled, and $a$ is the acceleration. In equation 4.10b, $m$ is the mass, and $F$ is the force. The scaling law for the response time is now easily determined:

$$t' = \sqrt{\frac{2(sx)(s^3m)}{s^f}}$$
$$t' = s^{\frac{4-f}{2}}t \tag{4.11}$$

The scaling law for the force, $s^f$, has been left generic. From equation 4.11, you can quickly determine how the response time will scale once you know the force's scaling law. For example, the magnetic force will lead to response times scaling $t \propto s^0$ for constant current density and $t \propto s^1$ for constant temperature rise.

Ideally, the exponent for the scaling law of a response time is greater than zero. This indicates that the response time decreases as the system is the system is miniaturized. As can be seen from equation 4.11, the force scaling law, $s^f$, must be very bad for this to be the case.

### 4.1.5 Power

Another important system parameter is power. This can be calculated using the following equation:

$$P = \frac{Fx}{t} \tag{4.12}$$

In equation 4.12, $P$ is the power consumed by accelerating the device, and the other variables have already been introduced. It follows that:

$$P' = \frac{(s^f F)(sx)}{s^{\frac{4-f}{2}}t}$$
$$P' = s^{\frac{3}{2}f-1}P \tag{4.13}$$

In addition to the power consumed per device (equation 4.13), it is also important to know the power consumed per unit volume.

$$\left(\frac{P}{V}\right)' = s^{\frac{3}{2}f-4}\left(\frac{P}{V}\right) \tag{4.14}$$

### 4.1.6 Temperature

If lumped parameter modeling of a device's temperature is justified[3], then the temperature of the device follows the following relation [5]:

$$\rho c L^3 \frac{\mathrm{d}T}{\mathrm{d}t} = -hL^2\left(T - T_\infty\right) \tag{4.15}$$

In equation 4.15, $\rho$ is the density of the device, $c$ is the device's specific heat, $L$ is the characteristic length of the device, $h$ is the convective heat transfer coefficient, $T$ is the device's temperature, and $T_\infty$ is the ambient temperature. The left-hand side of equation 4.15 expresses the power available through change in temperature, while the right-hand side expresses the power flow due to convection.

The solution of equation 4.15 is simple. The result shows that the temperature decays exponentially to the ambient temperature with a time constant equal to the following equation:

$$\tau_{\text{thermal}} = \frac{\rho c L}{h} \tag{4.16}$$

The scaling of the thermal time constant with size thus goes as:

$$\tau'_{\text{thermal}} = s\tau_{\text{thermal}} \tag{4.17}$$

Thus, one should expect the temperature of miniature devices to reach equilibrium very quickly. This is due to the fact that a device's heat capacity is proportional to its volume ($s^3$), while the rate of heat transfer is proportional to its surface area ($s^2$). Thus, the heat capacity decreases faster than that rate of heat conduction.

Because of the above, it is possible to create large temperatures differences with very little power. The main reason is that the rate of heat transfer scales as $s^2$ [3]. The amount of power required to maintain a temperature thus scales very favourably.

## 4.2 Casimir Force

Even a perfect vacuum is not empty. The Casimir force is the result of quantum fluctuations that exist in a vacuum even at absolute zero. Placing a body in the vacuum alters the quantum fluctuations, the result of which is the Casimir effect.

The Casimir effect is most often portrayed as creating a force that occurs between close parallel conducting plates. However, in general, all material bodies alter the zero-point fields around and between them. Therefore, all bodies in proximity will develop a Casimir force.

Still, parallel plates remain the classic example of the Casimir effect. The Casimir force between two parallel plates can be calculated using the following equation [6–8]:

$$F_c = \frac{\Re \eta A}{d^4} \tag{4.18}$$

$$\mathfrak{R} \equiv \frac{\hbar c \pi^2}{240} \tag{4.19}$$

In equation 4.18, $F_c$ is the Casimir force, $\eta$ is a materials factor, $A$ is the area of the plates, and $d$ is the distance separating the plates. The constant, $\mathfrak{R}$ is defined in equation 4.19, where $\hbar$ and $c$ have their usual physical meanings. If the plates are not perfectly parallel, then equation 4.18 is only an approximation. However, the error is less than $1\%$ for angles up to 5.7° [8].

The materials factor, $\eta$, depends on the dielectric permittivities of the plates and the intervening material. The maximum value for $\eta$ occurs for perfectly conducting plates separated by vacuum, for which $\eta = 1$. More details can be found in [6].

Although the Casimir force is very weak, its importance increases as devices are miniaturized. Whereas most forces decrease in strength as systems are shrunk, the Casimir effect benefits because of its extreme sensitivity to the separating distance ($F_c \propto d^{-4}$).

$$\begin{aligned} F_c' &= \frac{\mathfrak{R}\eta(s^2 A)}{(sd)^4} \\ &= s^{-2} F_c \end{aligned} \tag{4.20}$$

Therefore, the importance of Casimir forces increases as devices are miniaturized.

## 4.3 Surface Forces

As devices are miniaturized, the ratio of their volume compared to their surface area changes as well (equation 4.21b). As this ratio shrinks, the relative importance of surface forces increases.

$$\left(\frac{V}{A}\right)' = \left(\frac{s^3 V}{s^2 A}\right) \tag{4.21a}$$

$$\left(\frac{V}{A}\right)' = s\left(\frac{V}{A}\right) \tag{4.21b}$$

### 4.3.1 Electrostatic Force

Electrostatic attraction between surfaces can be caused either through charging of the surface or due to differing work functions of the surfaces [9].

Charging is not often a problem for conducting materials, since charging can be easily controlled. However, charges can also accumulate on the surface of insulating materials. These charges are very difficult to remove, and can lead to electrostatic forces of attraction. Even when a material's bulk is conducting, the surface may be insulating[4].

As previously covered, electrostatic forces scale well. Thus, unwanted forces due to surface charging can be a problem in microscale devices.

Differing work functions will also result in electrostatic attraction. If there are no charges, then the electrostatic force can be calculated using the difference in work functions as the potential difference between the two bodies.

### 4.3.2 van der Waals Force

On average, the electrons that surround the nucleus of an atom are symmetrically distributed. However, random fluctuations of the electrons can lead to the atom exhibiting a dipole. These dipoles create short-term electric fields that influence the surrounding atoms. The instantaneous electric fields can induce dipoles that lead to an attractive force between the atoms.

For two parallel surfaces that are very close together ($d_0 \lesssim 20$ Å), the force per unit area can be determined as follows [9]:

$$P_{\mathrm{vdW}} = \frac{\mathfrak{A}}{6\pi d^3} \tag{4.22}$$

In equation 4.22, $P_{\mathrm{vdW}}$ refers to the van der Walls pressure, $\mathfrak{A}$ is the Hamaker constant[5], and $d$ is the distance of separation. This equation applies to the non-retarded regime[6].

For larger separations, the retarded regime, the two instantaneous dipoles no longer interact as strongly. The force then becomes [9]:

$$P_{\mathrm{vdW}} = \frac{\mathfrak{A} d_0}{6\pi d^3 \left(d + d_0\right)} \tag{4.23}$$

Equation 4.23 introduced the variable $d_0$, the characteristic distance.

### 4.3.3 Hydrogen Bonding

The bulk of a crystalline material is made up of regularly spaced atoms. However, at the surface of the crystal, the atoms have additional unused bonding sites, which are chemically active.

Depending on chemical treatment, different chemical groups may bond to the surface. The most common situation is for these sites to be terminated by either a single hydrogen atom (-H) or an hydroxyl group (-OH). In particular, the surface atoms of hydrophilic silicon often bond to hydroxyl groups [9].

In both cases, the outer surface of the material is mostly hydrogen atoms. Especially when connected to oxygen atoms, the hydrogen atoms do not maintain strong possession of their electron and so present a positive charge. This positive charge can then induce dipoles in surrounding atoms, creating attractive electrostatic forces.

### 4.3.4 Contact Forces

The above forces discussed materials separated by either vacuum or a fluid (typically air). However, the materials eventually come into contact. If the materials are solid, then contact forces occur.

Surfaces are not perfectly flat. Some parts, called asperities, are higher than others, and the asperities come into contact first. Repulsive interatomic forces then exist as the atoms from the previously separate materials come into proximity. As increased pressure is put on the solid-solid interface, the asperities deform elastically allowing the surfaces to draw closer.

It is often important to distinguish between the apparent contact area and the real contact area. The apparent contact area is the macroscopic or mesoscopic measure of the area. For example, two plates measuring $1\text{mm}^2$ have an apparent contact area of $1\text{mm}^2$. However, because of asperities, not all of the surface is in contact. The real contact area is the measure of the microscopic area of the interface where the two bodies are actually in contact.

### 4.3.5 Adhesion

Creating an interface between different materials, for example between water and air, takes energy. For example, when two water droplets touch they join together into a single droplet because they can then minimize their surface area, and therefore their surface energy.

Similarly, solids like to minimize their surface area as well. However, since they don't flow, it is less common to see solid merging like droplets. However, surface-micromachined devices have large, relatively flat, surfaces separated by small distances. It often takes less energy for surface-micromachined structures to elastically deform, than to maintain all of their surfaces.

Calculations of adhesion will be covered in chapter 5.

### 4.3.6 Capillary Forces and Solid Bridging

The existence of a liquid between two surfaces can create a force of attraction called the capillary force. When the structures are fully immersed, no force exists. However, as the liquid evaporates, the water shrinks into water droplets. If the required surface energy for the liquid is less with the structural material than with air, the liquid droplets will preferentially form underneath the surface-micromachined structures.

As they shrink into droplets, the internal pressure of the droplet also drops. If the top surface of a micromachined device is exposed to air, this creates a net force pushing the device to the wafer surface. Once in contact with the surface, adhesion can occur.

Solid bridging is another important consequence of drying. As a droplet dries, any dissolved impurities are concentrated as the droplet shrinks. Finally, the impurities can result in a solid residue [9]. This solid residue can then form a solid compound bridging two otherwise separate bodies.

## 4.4 Friction

In macroscale devices, the force of friction between two surfaces is relatively easy to compute. Friction is linearly proportional to the normal force:

$$F_f = \mu_s F_N \tag{4.24}$$

In equation 4.24, $F_f$ is the force of friction, $\mu_s$ is the coefficient of static friction, and $F_N$ is the normal force. Once the two surface are in motion relative to one another, the coefficient of static friction is replaced with the coefficient of dynamic friction, $\mu_d$.

In microscale devices, designers need to be more careful. The roughness of devices' surfaces may be sufficient such that equation 4.24 no longer applies. This is in part indicated by several measured values for $\mu_s$. In macroscale devices, $\mu_s \in (0, 1)$. However, tests of static friction in microdevices have measured coefficients of static friction as high as $4.9 \pm 1$.

Two source friction can be identified [10]: ploughing of asperities and shearing at the interface. The first case typically applies when one material is significantly harder than the other. The high points of the hard material, asperities, plough through and deform the softer material. In micromachined devices, however, shearing of junctions is more common. The surfaces of micromachined devices are usually equally hard, and the asperities are not very sharp.

When the asperities are blunt, the asperities will undergo elastic deformation when the surfaces are forced together. One then has to distinguish between the apparent, or macroscopic, area in contact, and the real contact area. The real contact area refers to actual interfacial area where interatomic forces are preventing the two surfaces from passing through each other. The shearing friction in this case is [10]:

$$F_f = sA_r \tag{4.25}$$

In equation 4.25, the frictional force is determined by the product of the shear strength of the interface, $s$, and the real contact area, $A_r$. While in general, $s$ may depend on pressure, we will ignore this case. Depending on the assumptions made about the surface[7], different equations relating friction and the applied normal force can be determined. If the asperities deform plastically, one finds that equation 4.24 still applies with $\mu_s = s/H$, where $H$ is the composite hardness. If the asperities deform elastically, and are spherical, then equation 4.24 still applies, but with $\mu_s = s/p_e$, where $p_e$ is the mean elastic contact pressure.

In practice, determining the force of friction is very difficult. For example, the coefficient of friction depends heavily on the environment [11]. Not just the present conditions are important, but past conditions as well, since gas molecules can be adsorbed onto the surface, which alters the coefficient of friction.

Monolayer coatings can reduce the friction in part because they have lower surface energies, and so reduce the shear strength of the interface. However, since they are thin, they do not modify the elastic properties of the device ($p_e$ constant).

Despite the above, there are two major deviations from equation 4.25 that designers must contend with: the contribution of the adhesion force and the case where few asperities make contact.

### 4.4.1 Adhesion Force

In equation 4.24, the force of friction, $F_f$, varies proportionally to the applied normal force, $F_N$. In reality, adhesive forces between the two surfaces also load the interface. Calculation of the friction force should then follow [10]:

$$F_f = \mu_s (F_a + F_N) \tag{4.26}$$

Above, the force of adhesion, $F_a$ had been added. Clearly, when the normal force is much greater than the adhesion force, then equation 4.26 simplifies to equation 4.24. Unfortunately, $F_a$ may be several micro-Newtons, and so cannot be ignored in micromachined devices.

### 4.4.2 Limited Contact

All of the above are based on statistical models. When the applied normal force is not very large, then the number of asperities in contact may become very small, and the statistical models breakdown. Particularly, even at very low loads, the number of asperities in contact will remain at 1, 2, or 3, depending on the number of degrees of freedom of the surface [10]. Different equations for the force of friction are then required. Maintaining the assumption that shearing at the interface is the dominant mechanism, then equation 4.25 still holds as long as one realizes that $A_r$ is a function of the applied normal force [10]. This leads to:

$$F_f = sN\pi R \left[ \frac{D}{N\sqrt{R}} \left( F_N + 2N\pi w_a R \right) \right]^{2/3} \tag{4.27}$$

Above, the new variables are: $N$ is the number of equally loaded asperities, $R$ is the radius of the asperities, and $w_a$ is the work of adhesion.

## 4.5 Continuum Mechanics Approximation

Many theories rely on the approximation that the subject of interest is continuous, when in fact the subject is discrete. For example, fluid mechanics often works from the idea that a fluid is a continuous body which can be characterized by pressure, density, viscosity, etc. However, a fluid is in reality made up of many discrete molecules. Material properties, such as pressure, are statistical properties of the fluid's molecules. Friction between surfaces in limited contact (section 4.4.2) is another example.

When the space and time scales are sufficiently large, these statistical properties are very useful. Unfortunately, at microscopic scales, many of these approximations weaken. Designers must therefore be aware that the many theories that depend on continuum mechanics may break down in micromachined devices.

### 4.5.1 Material Properties

Transport properties of materials are known to exhibit size effects [3, 12]. Electrical and thermal conduction, for example, are influenced by the material surface and grain boundaries. Once a film's thickness reaches the mean free path of carriers[8], the film's conductivity will decrease significantly.

Typically, grain boundaries increase both thermal and electrical resistivity. Hence, the electrical and thermal conductivity of polycrystalline materials should be lower than single crystalline materials. In large samples, the effects of grain boundaries are averaged out. As dimensions approach the grain size, carriers may find regions where they do not need to cross any grain boundaries. This would decrease the local resistivity.

The yield strength of brittle materials is also sensitive to size [3, 12]. Fracture is initiated by pre-existing flaws on the material's surface or within the material's bulk. Smaller devices lower the probability that a sufficiently large flaw will be present, and so exhibit higher yield strengths. This is especially true as device volumes approach the defect density.

In addition to changes in the average of material properties, designers should expect increased variation in material properties [3]. Local variations in grain size, which are averaged out in larger devices, will be reflected in variations in the material properties, even for devices produced together.

### 4.5.2 Mean Free Path

In a gas, the molecules are free to move about since they do not interact with one other strongly. However, they will still collide. The mean free path is the average distance between collisions for a molecule in a gas.

The average distance a gas molecule travels between collisions is based on three factors: its cross-sectional area, its speed, and the density of molecules. The mean free path can be calculated as follows [13]:

$$\lambda = \frac{1}{\sqrt{2}\pi d^2 n_v} \tag{4.28a}$$

$$\lambda = \frac{RT}{\sqrt{2}\pi d^2 N_A P} \tag{4.28b}$$

In equation 4.28a, $\lambda$ is the mean free path, $d$ is the diameter of the molecule, and $n_v$ is the number of molecules per unit volume. In equation 4.28b, the Ideal gas law has been used to calculate $n_v$. The additional variables are: $R$, the universal gas constant[9]; $T$, the absolute temperature; $N_A$, Avogadro's number[10]; and $P$, the pressure.

Knowledge of the mean free path is important when it approaches the system's dimensions. As system dimensions approach the mean free path, continuum mechanics approximations fail. For example, the thermal conductivity of air decreases when the system dimensions approach the mean free path.

Whether or not continuum mechanics will hold for a fluid can be checked, in part, but determining the Knudsen number (section 2.3.3). When $Kn$ is less than 0.1, continuum mechanics should hold. When $Kn$ is between 0.1 and 1.0, continuum mechanics may still hold, although new effects may be present[11].

### 4.5.3 Paschen's Curve

One particularly imporant break down in continuum mechanics is Paschen's curve, which relates the breakdown voltage of air between two plates to the plate separation [14, 15]. For macroscopic distances, the breakdown voltage can be calculated simply by knowing the breakdown field strength[12]. However, as can be seen from figure 4.2, this relationship does not hold for small distances.

As also can be seen from figure 4.2, a minimum breakdown voltage exists. For smaller electrode gaps, insufficient collisions occur to maintain the avalanche effect. In air, the minimum breakdown voltage is between 330 and 340 volts, and occurs between 5 and 6 $\mu$m.

Paschen's curve is extremely important for micromachined structures because, if continuum mechanics held, it would be impossible to apply a measurable voltage to closely separated objects. At small distances, even small voltages will lead to strong electric fields and dielectric breakdown.

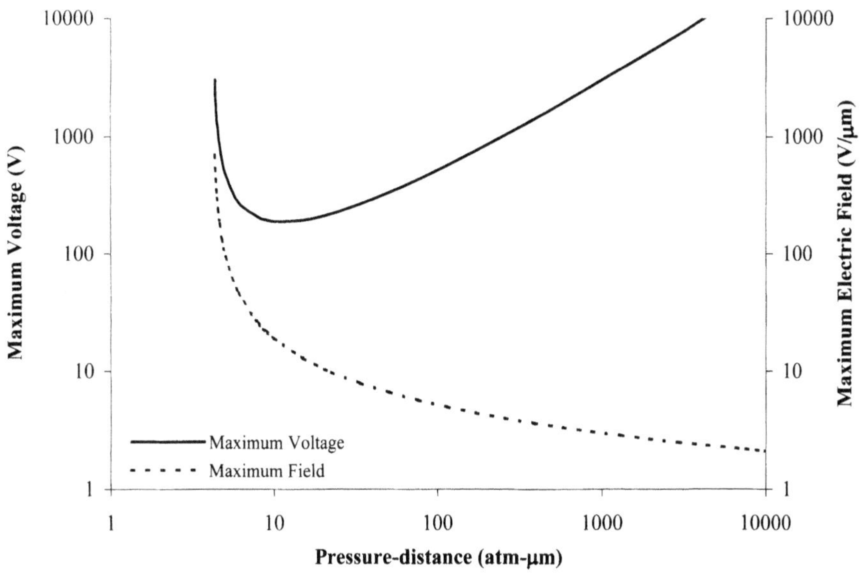


*Figure 4.2.* Breakdown voltage and breakdown electric field vs. plate separation in air.

Paschen's curve, along with other scaling effects, leads to the preference for electrostatic over magnetostatic actuation in most microdevices [16].

The following derivation is taken in large part from Judy [15].

The breakdown of a dielectric, such as air, occurs because of impact ionization. The number of new free electrons generated by ionization in a short distance can be determined using the following equation:

$$dn = n\alpha \mathrm{d}x \tag{4.29}$$

Above, $n$ is the current number of electrons, $\alpha$ is the ionization coefficient, and $\mathrm{d}x$ is the distance traveled. An electron falling a distance $d$, will generate $e^{\alpha d}$ electrons and $e^{\alpha d} - 1$ positive ions through impact ionization.

To generate electrical breakdown the supply of electrons needs to be maintained. New free electrons are created when the positive ions reach, and strike, the cathode. If the probability that a positive ion will generate a new free electron when it reaches the cathode is $\gamma$, then the following condition needs to be maintained for breakdown:

$$\gamma \left( e^{\alpha d} - 1 \right) \geq 1 \tag{4.30}$$

In addition to the impact ionization caused by the positive ions, dielectric breakdown also depends on the impact ionization of electrons. Paschen's curve results from two assumptions: electrons lose all of their kinetic energy on impact and electrons ionize as soon as their kinetic energy is more than the ionization potential of the gas. Electrons have a probability of traveling a distance based on the mean free path (equation 4.28a).

$$p = \exp\left(\frac{-x_{\text{ionization}}}{\lambda}\right) \tag{4.31a}$$

$$p = \exp\left(\frac{-V_{\text{ionization}}}{E\lambda}\right) \tag{4.31b}$$

Using the above, $p$ is the probability that an electron will gain sufficient kinetic energy to ionize the gas molecule it collides with, $x_{\text{ionization}}$ is the distance that the electron must travel, $\lambda$ is the mean free path, $V_{\text{ionization}}$ is the ionization potential of the gas molecules, and $E$ is the strength of the electric field. Additionally, there are $1/\lambda$ impacts per unit distance. This leads to the following equation for the ionization coefficient:

$$\alpha = \frac{1}{\lambda}\exp\left(\frac{-V_{\text{ionizaton}}}{E\lambda}\right) \tag{4.32}$$

At breakdown, equations 4.30 and 4.32 must both be true. Substituting $E = V_{max}/d$, we get the following equation for the breakdown voltage:

$$V_{max} = \frac{V_{\text{ionization}}d}{-\lambda \ln\left(\frac{\lambda}{d}\ln\left(1+\frac{1}{\gamma}\right)\right)} \tag{4.33}$$

The above allows people to predict the breakdown voltage. However, clarifying the importance of the result for micromachining requires substituting in for the mean free path, $\lambda$.

$$V_{max} = \frac{Apd}{\ln(pd)+B} \tag{4.34a}$$

$$A = \frac{V_{\text{ionization}}}{1.013\times 10^5\text{Pa}\quad \lambda_0} \tag{4.34b}$$

$$B = -\ln\left(1.013\times 10^5\text{Pa}\quad \lambda_0 \ln\left(1+\frac{1}{\gamma}\right)\right) \tag{4.34c}$$

Above, $p$ is the pressure in atmospheres, $d$ is the distance between the electrodes, and $\lambda_0$ is the mean free path at one atmosphere.

## Notes

1 Constant current scaling might result if the current density is limited by electromigration.

2 The insulator will have a maximum temperature due to melting, but other limits may exists.

3 A lumped parameter model is accurate if the Biot number is less than 0.1. The Biot number can be calculated using $Bi = hL/\kappa$, where $h$ is the convective heat transfer coefficient, $L$ is the characteristic length, and $\kappa$ is the device material's thermal conductivity.

4 For example, doped polysilicon is conducting. However, the surface may be insulating due to a thin oxide layer or through surface treatments [9].

5 $\mathfrak{A} \approx 1.6\text{eV}$ for bare silicon.

6 In the non-retarded regime, changes in the electric field propagate between the interacting dipoles faster than the dipoles fluctuate.

7 These assumptions include assumptions about the shape, size, and height of the asperities.

8 Electrons for electrical conductivity and phonons for thermal conductivity.

9 8.3145 J/mol °K

10 $6.0221 \times 10^{23}$

11 This is often called the slip-flow regime, relating to changing in boundary conditions.

12 At one atmosphere, the breakdown voltage is $E_{\mathrm{cr}} = 3 \times 10^6 \mathrm{Vm}^{-1}$.

## References


[1] C. Liu and Y. Bar-Cohen. Scaling laws of microactuators and potential applications of electroactive polymers in mems. In *Proceedings of SPIE's 6th Annual International Symposium on Smart Structures and Materials*, pages 345–354. SPIE, 1999.

[2] R.R.A. Syms. Scaling laws for mems mirror-rotation optical cross connect switches. *Journal of Lightwave Technology*, 20(7):1084–1094, 2002.

[3] J.W. Judy. Microelectromechanical systems (mems): fabrication, design and applications. *Journal of Smart Materials and Structures*, 10(6):1115–1134, 2001.

[4] W.S.N. Trimmer. Microrobots and micromechanical systems. *Sensors and Actuators*, 19(3):267–287, 1989.

[5] M. Gad el Hak. The fluid mechanics of microdevices – the freeman scholar lecture. *Journal of Fluids Engineering*, 121(1):5–33, 1999.

[6] E. Elizalde and A. Romeo. Essentials of the casimir effect and its computation. *American Journal of Physics*, 59(8):711–719, 1991.

[7] P.W. Milonni and M.-L. Shih. Casimir forces. *Contemporary Physics*, 33(5):313–322, 1992.

[8] F.M. Serry, D. Walliser, and G.J. Maclay. The role of the casimir effect in the static deflection and stiction of membrane strips in microelectromechanical systems (mems). *Journal of Applied Physics*, 84(5):2501–2506, 1998.

[9] R. Maboudian and R.T. Howe. Critical review: Adhesion in surface micromechanical structures. *Journal of Vacuum Science and Technology B*, 5(1):1–20, 1997.

[10] N.R. Tas, C. Gui, and M. Elwenspoek. Static friction in elastic adhesion contacts in mems. *Journal of Adhesion Science and Technology*, 17(4):547–561, 2003.

[11] K. Deng and W.H. Ko. A study of static friction between silicon and silicon compounds. *Journal of Micromechanics and Microengineering*, 2(1):14–20, 1992.

[12] J.N. Ding, Y.G. Meng, and S.Z. Wen. Specimen size effect on mechanical properties of polysilicon microcantilever beams measured by deflection using a nanoindenter. *Materials Science and Engineering B*, 83(1):42–47, 2001.

[13] R. Serway. *Physics for Scientists and Engineers with Modern Physics*. Saunders College Publishing, Philadelphia, 3 edition, 1990.

[14] A.R. von Hippel. *Molecular Science and Molecular Engineering*. The M.I.T. Press, New York, 1959.

[15] J.W. Judy. *Magnetic Microactuators with Polysilicon Flexures*. PhD thesis, University of California at Berkeley, 1994.

[16] S.F. Bart, T.A. Lober, R.T. Howe, J.H. Lang, and M.F. Schlecht. Design considerations for micromachined electric actuators. *Sensors and Actuators*, 14(3):269–292, 1988.

# Chapter 5

# MECHANICS

Surface micromachining is, after all, about mechanical structures. This chapter does not aspire to replace a proper foundation in mechanical design. However, at minimum, a grounding in structural mechanics and beams is necessary. This chapter will provide a brief introduction to the subject. More dedicated readers are referred to Timoshenko and Gere [1], a standard introduction to mechanics.

Singly supported and doubly supported beams are very common in surface-micromachining. These two structures in particular will be reviewed. Their susceptibility to adhesion, although not a traditional topic in mechanics, is also covered.

Finally, the topic of pseudo-rigid body modelling will be covered. Pseudo-rigid-body modelling is a method to simplify compliant mechanical systems to a more traditional rigid-body model, where all of the components are perfectly rigid.

A pseudo-rigid model captures the behaviour a compliant mechanism by placing springs at the joints of a rigid-body model. To begin, a rigid-body model mimics the range of motion is selected. Then, the distributed compliance of the mechanism is collected into lumped springs which are collocated with the joints.

## 5.1 Beam Equations

Before discussing the two specific examples of beams this chapter will cover, a general introduction into beam theory will be given.

### 5.1.1 Axial Loading

By definition, the strain is defined by $\epsilon = \Delta L/L$ [1], and is a dimensionless parameter describing the deformation of the structure. The stress is defined as the force carried per unit area, and so can be determined from the equation $\sigma = F/A$. If the deformation is linearly related to the applied stress (Hooke's law), then the stress and strain are related by [1]:

$$\sigma = E\epsilon \tag{5.1}$$

In equation 5.1, $\sigma$ is the stress, $E$ is Young's modulus, and $\epsilon$ is the strain. Young's modulus, which is also known as the modulus of elasticity, is similar to the compliance of a spring (section 10.2).

When the beam is loaded axially, transverse dimensional changes also occur. For example, a bar in tension is longer, but also thinner. The degree of transverse dimensional change is known as Poisson's ratio, $\nu$, and takes a value between 0 and 0.5. However, empirical values are typically in the range of 0.25 to 0.35 [1].

The deflection of a beam under axial loading is a simple question for the mechanics of material, and can be derived from equation 5.1, and leads to the following equation [1]:

$$\Delta L = \frac{FL}{EA} \tag{5.2}$$

In equation 5.2, the deflection, $\Delta L$, is determined using the applied axial force, $F$, the beam's length, $L$, Young's modulus of the beam's material, $E$, and the beam's cross-sectional area, $A$.

### 5.1.2 Transverse Loading

Many beams can be described by the Euler-Bernoulli beam equations, the derivation of which begins with understanding how a differential beam element carries a transverse load.

Transverse loads create shear stress in the beam. Additionally, static equilibrium of the transverse forces acting on the beam requires that moments also exist. Moments, whether they are due to an applied load or are a reaction force, must be carried by the beam (figure 5.1).

Analysis of a loaded beam starts with the two equations necessary for equilibrium (figure 5.1):

$$\frac{\mathrm{d}V}{\mathrm{d}x} = -p(x) \tag{5.3a}$$

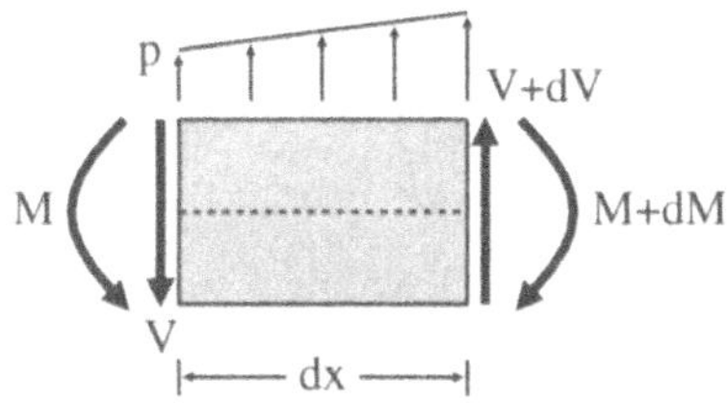


*Figure 5.1.* Forces and moments acting on an infinitesimal length of beam.

$$\frac{\mathrm{d}M}{\mathrm{d}x} = V(x) \tag{5.3b}$$

In the above equations, $V$ is the shear load carried by the beam, $x$ is the distance along the axial direction of the beam, $p$ is the distributed load, and $M$ is the bending moment carried by the beam. Equation 5.3a is due to force equilibrium in the transverse direction, and equation 5.3b is due to moment equilibrium. Combining these two equations, one can relate the bending moment to the distributed load:

$$\frac{\mathrm{d}^2 M}{\mathrm{d}x^2} = -p(x) \tag{5.4}$$

The bending moment, $M$, is a function of the local stresses over the cross-sectional surface. This integral can be substituted into the previous equation. Assuming the beam is made from a single material, then one can also easily substitute the stress-strain relationship (equation 5.1) to get an expression based on the beam's stress rather then strain.

$$\frac{\mathrm{d}^2}{\mathrm{d}x^2}\left(\int\int y\sigma(y,z) \quad \mathrm{d}y\mathrm{d}z\right) = -p(x) \tag{5.5a}$$

$$\frac{\mathrm{d}^2}{\mathrm{d}x^2}\left(E\int\int y\epsilon(y,z) \quad \mathrm{d}y\mathrm{d}z\right) = -p(x) \tag{5.5b}$$

The integral in equations 5.5a and 5.5b are either over the stress or over the strain. However, the strain across the a beam varies linearly[1]. The variation in strain is directly related to the radius of curvature. In the paraxial approximation, the radius of curvature can be approximated by the second derivative of the transverse displacement. This leads to:

$$\frac{\mathrm{d}^2}{\mathrm{d}x^2}\left(E\int\int y\left(y\frac{\mathrm{d}^2 w}{\mathrm{d}x^2}\right) \quad \mathrm{d}y\mathrm{d}z\right) = -p(x) \tag{5.6a}$$

$$\frac{\mathrm{d}^2}{\mathrm{d}x^2}\left(E\frac{\mathrm{d}^2 w}{\mathrm{d}x^2}\int\int y^2 \quad \mathrm{d}y\mathrm{d}z\right) = -p(x) \tag{5.6b}$$

Finally, the double integration is, by definition, the area moment of inertia.

$$\frac{\mathrm{d}^2}{\mathrm{d}x^2}\left(EI\frac{\mathrm{d}^2 w}{\mathrm{d}x^2}\right) = -p(x) \tag{5.7}$$

Equation 5.7 is the Euler-Bernoulli beam equation, in which $E$ is Young's modulus, $I$ is the area moment of inertia, $w$ is the beam's transverse displacement, $x$ is the distance along the beam's axis, and $p$ is the transverse load (force per unit length).

If the transverse load is zero along the entire length of the beam, then the beam is either unloaded or in pure bending. In pure bending, the only loads are moments.

Often, the beam will be made from a single material and have a uniform cross-section. In this case, equation 5.7 simplifies to the following form:

$$EI\frac{\mathrm{d}^4 w}{\mathrm{d}x^4} = -p(x) \tag{5.8}$$

For rectangular beams, Young's modulus needs to be modified if the beam is 'wide.' The beam is considered wide when $w \geq 5t$, where $t$ is the beam thickness [2]. In this case, the biaxial Young's modulus must be used, which is $E/(1-\nu^2)$. Narrow beams simply use Young's modulus, $E$. Since Poisson's ratio, $\nu$, is in the range of $0$ to $0.5$, neglecting Poisson's ratio for wide beams does not lead to extremely large errors, and is easily overlooked.

If the beam is doubly supported, then the beam may also be carrying an axial load. Furthermore, even if the beam is not initially stressed axially, when the beam deforms it can carry some of the transverse load axially. This leads to stress stiffening. If axial loads need to be accounted for, an additional term needs to be added to equation 5.8, as follows [2]:

$$EI\frac{\mathrm{d}^4 w}{\mathrm{d}x^4} - T\frac{\mathrm{d}^2 g}{\mathrm{d}x^2} = -p(x) \tag{5.9}$$

Above, we have added the variable $T$, which is the axial force carried by the beam. Note that the axial load on the beam can be determined by the average stress and cross-sectional area, $T = \bar{\sigma}A$.

## 5.2 Torsion

In addition to axial and transverse loading[2], beams can resist applied moments. When the axis of rotation is aligned with the beam's axis, then the applied moment will twist the beam. If the beam has a uniform cross-section, and the entire beam experiences the same moment, then the bar is in pure torsion.

In pure torsion, the axis of the beam remains straight, and the beam does not change length. However, the cross-sections of the beam are rotated about the beam's axis. The amount of rotation varies linearly along the length of the beam. Unfortunately, this ideal only applies to beams with circular cross-sections, a situation that is not common in surface-micromachining.

When the beam does not have a circular cross-section, torsion leads to warping. The cross-sections of the beam do not remain planar after the moment is applied.

Under pure torsion, the beam resists the applied load using pure shear. For a circular cross-section, this allows the maximum shear strain to be easily calculated, and from there the maximum shear stress:

$$\gamma_{\text{max}} = \frac{r\theta}{L} \tag{5.10a}$$

$$\tau_{\text{max}} = \frac{Gr\theta}{L} \tag{5.10b}$$

In equation 5.10a, $\gamma_{\text{max}}$ is the maximum shear strain, $r$ is the beam's radius, $\theta$ is the beam's total rotation, and $L$ is the beam's length. Equation 5.10b introduces $G$, the shear modulus.

The maximum shear stress can also be determined from the applied torque:

$$\tau_{\text{max}} = \frac{rT}{I_p} \tag{5.11}$$

In equation 5.11, the maximum shear stress is calculated based on previously defined quantities and the applied moment or torque, $T$, and the polar moment of inertia, $I_p$. For circular beams, $I_p = \pi r^4/2$.

From the previous equation, one can easily see that the rotation can be found directly from the applied moment using the following equation:

$$\theta = \frac{TL}{GI_p} \tag{5.12}$$

In surface-micromachining, beams do not often have circular cross-sections. Rectangular cross-sections are common. In this case, the polar moment of inertia is:

$$I_p = \frac{w^3t + wt^3}{12} \tag{5.13}$$

Above, $w$ is the beam's width and $t$ is the beam's thickness. However, note that a rectangular beam will not be in pure torsion. The equations provided in this section are therefore only an approximation.

*Figure 5.2.* Illustration of a singly supported beam (cantilever).

## 5.3 Cantilevers

A cantilever is a beam that is fully supported on one end (figure 5.2). In two dimensions, this means that the supported end's position is fully fixed[3]

The details of beam deformation under various loading configuration can be determined from the Euler-Bernoulli beam equation (equation 5.7). Also, the equation for most load conditions can also be looked up in appropriate hand-books [3].

### 5.3.1 Deflection under End Loads

A cantilever acting under an end load is a useful model for many beam problems where the beam is connected only at the ends. The reaction forces are easily determined from static equilibrium conditions. The deflections are also important.

Under *axial load*, the deflection is determined straightforwardly from the strength of materials. Up to the elastic limit of the material, the deflection is linearly proportional to the beam's length, $L$, and inversely proportional to its cross-sectional area, $A$, and elasticity, $E$.

$$w_x = \frac{F_x L}{AE} \tag{5.14}$$

Under a *transverse load*, the Euler-Bernoulli beam equation can be used with the applied load represented as an impulse located at the end of the beam. The boundary conditions, in addition to setting $y$ and $\theta$ at the fixed end of the beam, also include setting the bending moment to zero at the free end of the beam.

$$p(x) = \delta(x - L) \tag{5.15}$$

The displacement is then given by the following equation. Note that the maximum deflection occurs at the beam end, so the maximum deflection can then be found evaluating $w_y(L)$.

$$w_y(x) = \frac{-F_y x^3 (3L - x)}{6EI} \tag{5.16}$$

$$w_{max} = \frac{-F_y L^3}{3EI} \tag{5.17}$$

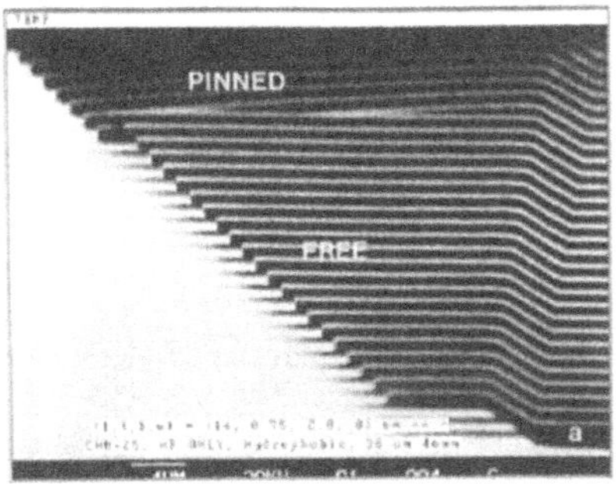


*Figure 5.3.* SEM of micromachined polysilicon cantilever beams of increasing length [4]. Note that the longer cantilevers are stuck to substrate surface, while the shorter cantilevers are not. Copyright © 1997 Kluwer.

Under a *moment load*, the displacement is given by:

$$w_y(x) = \frac{-Mx^2}{2EI} \quad (5.18)$$

$$w_{max} = \frac{-ML^2}{2EI} \quad (5.19)$$

## 5.3.2 Adhesion

Whether or not a cantilever will stick to the substrate when removed from a liquid can be determined by the elastocapillary number, $N_{ec}$. The elastocapillary number is the ratio between the elastic forces necessary to deform the cantilever into contact with the substrate and the adhesion forces necessary to maintain contact [4–6]. If the elastocapillary number is greater than one, the cantilever will be free [4–6].

$$N_{ec} = \frac{2EIh^2t^3}{9\gamma_l \cos\theta_c l^4 \left(1 + t/w\right)} \quad (5.20)$$

In equation 5.20, $E$ is the Young's modulus for the cantilever's material, $h$ is separation between the cantilever and the substrate, $t$ is the thickness of the cantilever, $\gamma_l$ is the surface tension of the liquid, $\theta_c$ is the contact angle, $l$ is the length of the cantilever, and $w$ is the width of the cantilever.

The transition point, $N_{ec} = 1$ is not sharp [4–6]. If the cantilever has sufficient agitation, it may shift. Thus, there is a finite probability that cantilevers will be free for $N_{ec} < 1$ and a finite probability that cantilevers will be stuck for $N_{ec} > 1$.

Strictly speaking, release processes (section 2.8) that avoid liquid drying do not need to worry about elastocapillary number. However, even if a surface-micromachined device finishes the release process without adhering, it may still touch the substrate at some point in the future. Therefore, ensuring that devices have $N_{ec} > 1$ will improve device reliability.

*Figure 5.4.* Illustration of a doubly supported beam (Euler column).

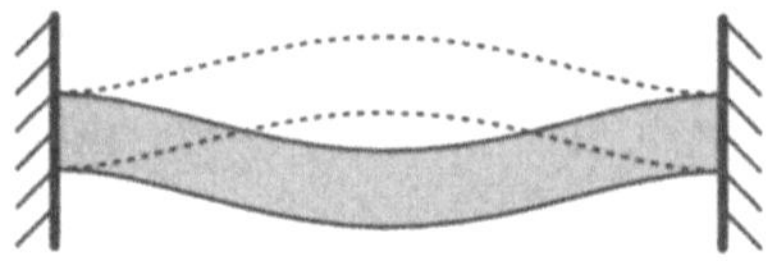

*Figure 5.5.* Illustration of a Euler-column that has buckled.

Similar to the elastocapillary number, there is the peeling bound, $N_p$. If the peeling bound is greater than one, then the force of adhesion is insufficient to keep the cantilever in contact with the substrate [4–6].

$$N_p = \frac{3Et^3h^2}{8\gamma_s l^4} \tag{5.21}$$

In equation 5.21, the variable $\gamma_s$ was introduced. This is the surface energy per unit area of the cantilever.

## 5.4 Euler-columns

A Euler-column is a beam which is fully fixed on both ends (figure 5.4). Deflection occurs in the middle of the beam. The deflection of Euler-columns under load can be analyzed using the Euler-Bernoulli beam equation with $w$ and $\mathrm{d}w/\mathrm{d}x$ fixed at both ends of the beam.

Since Euler-columns are fully supported at each end, end loads applied to the beam will not result in any deflection of the beam.

### 5.4.1 Buckling

A Euler-column may buckle if carrying a compressive load. Although forces applied to the ends of the beam will be carried by the supports, it is still possible for the beam to carry several forces. For example, a Euler-column may expand due to thermal expansion. The supports must then supply sufficient force to squeeze the beam back to its original length.

A Euler-column will be stable if the stress is tensile or below a critical value. For compressive loads above the critical stress, the Euler-column is unstable and will bukcle (figure 5.5). The critical strain and the critical stress can be determined as follows:

$$\epsilon_{CR} = \frac{\pi^2 t^2}{3L^2} \tag{5.22a}$$

$$\sigma_{CR} = \frac{E\pi^2 t^2}{3L^2} \qquad (5.22b)$$

In equation 5.22a, $\epsilon_{CR}$ is the critical buckling strain, $t$ is the film thickness, and $L$ is the unsupported length of the beam. Alternatively, one can use equation 5.22b, in which $\sigma_{CR}$ is the critical buckling stress, and $E$ is the modulus of elasticity. For the PolyMUMPs$^{TM}$ process, doubly supported beams as short as 210 $\mu$m may buckle.

### 5.4.2 Adhesion

Similar to the cantilever, one can use the elastocapillary number and the peeling bound to check whether adhesion will occur. For doubly supported beams, the elastocapillary number is given by [4–6]:

$$N_{ec} = \frac{128EIh^2t^3}{15\gamma_l \cos\theta_c l^4 (1 + t/w)} \left(1 + \frac{2\sigma_R l^2}{7Et^2}\right) \qquad (5.23)$$

Equation 5.23 introduced an additional variable, $\sigma_R$, to the elastocapillary number. The new variable is the residual uniform stress in the Euler-column (section 7.3.1). Tensile stress will stiffen the beam, making it more resilient to adhesion.

Also, the peeling bound must be determined [4–6]:

$$N_p = \frac{128Eh^2t^3}{5\gamma_s l^4} \qquad (5.24)$$

The variables used in equation 5.24 are the same as equation 5.21. The two equation are equal except for a constant scaling factor.

## 5.5 Pseudo-rigid body Modelling

As will become clear when we start discussing surface-micromachined devices, compliant systems (section 10.1) are important in the design of surface-micromachined devices. Compliant mechanical systems can be simplified for analytical modelling by using pseudo-rigid body modelling. In these models, elastic beams are replaced with a rigid-body models that approximates the motion of the elastic beams. Linear and torsional springs are added to the joints as required to recover the beams resistance to deformation.

There is a wide range of pseudo-rigid-body models [7]. The models are chosen not just on the compliant mechanism's structure, but also on the expected loading and the expected motion.

To illustrate the idea of pseudo-rigid body modelling, the small-length flexural pivot will be discussed. This is similar to the 'I'-joint, discussed in a later chapter (section 10.4.2).

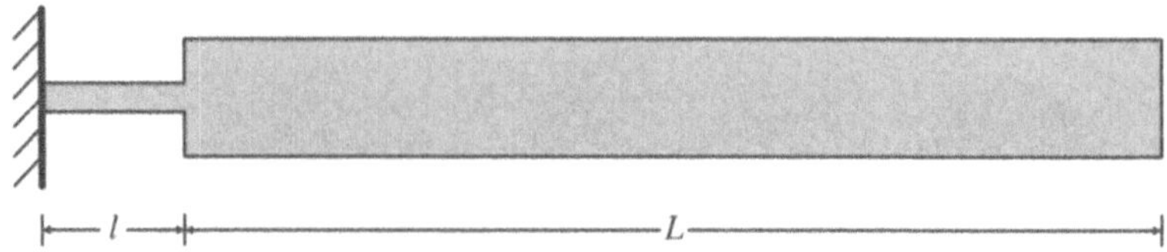


*Figure 5.6.* Illustration of a small-length flexural pivot [7].

A small-length flexural pivot is similar to a cantilever, but has two segments (figure 5.6). The first segment is short and flexible, and is used instead of a sliding contact revolute joint. The following approximations are used:

$$L \gg l \tag{5.25a}$$

$$(EI)_L \gg (EI)_l \tag{5.25b}$$

In equation 5.25a, $l$ is the length of the pivot, $L$ is the length of the rigid arm, and $(EI)_l$ and $(EI)_L$ refer to the bending stiffnesses of the respective segments.

The pseudo-rigid body approximation for the small-length flexural pivot replaces the compliant system as shown in figure 5.6 with a rigid body model. The compliant segment is replaced with a revolute joint that is located halfway along the flexure. Thus, if the rotational degree of freedom thus provided is denoted by $\theta$, the the position of the tip of the cantilever can be calculated as follows [7]:

$$x = \frac{l}{2} + \left(L + \frac{l}{2}\right)\cos\theta \tag{5.26a}$$

$$y = \left(L + \frac{l}{2}\right)\sin\theta \tag{5.26b}$$

To account for the pivot's stiffness, a torsional spring must be placed at the revolute joint. This stiffness can be calculated as follows[4] [7]:

$$k_\theta = \frac{(EI)_l}{l} \tag{5.27}$$

This pseudo-rigid model is most accurate when the cantilever is subject to pure bending[5]. However, a transverse load applied at the cantilever end will be supported as a bending moment in the small-length flexure. The deviation from pure bending can be assessed by determining the ratio of the bending moment at each end the small-length flexure. This ratio can be calculated as follows [7]:

$$\frac{M_1}{M_2} = 1 + \frac{l}{L} \quad (5.28)$$

Since the rigid segment should be much longer then the flexure, the error should be small. Calculations of the deflection of small-length flexural pivots is discussed in more detail by Howell [7].

## Notes

1 The strain is zero at the neutral axis.

2 This also includes moments applied to the beam where the axis of rotation is perpendicular to the beam's axis.

3 This includes $x$, $y$, and $\theta$.

4 This stiffness can be calculated easily from section 5.3.1.

5 i.e. the only load is an applied moment.

## References

[1] S.P. Timoshenko and J.M. Gere. *Mechanics of Materials.* Van Nostrand Reinhold, New York, 1972.

[2] P.M. Osterberg and S.D. Senturia. M-test: A testchip for mems material property measurement using electrostatically actuated test structures. *Journal of Microelectromechanical Systems*, 6(2):107–118, 1997.

[3] R.J. Roark. *Roark's Formulas for Stress and Strain.* McGraw-Hill, New York, 1989.

[4] C.H. Mastrangelo. Adhesion-related failure mechanisms in micromechanical devices. *Tribology Letters*, 3(3):223–238, 1997.

[5] C.H. Mastrangelo and C.H. Hsu. Mechanical stability and adhesion of microstructures under capillary forces – part i: Basic theory. *Journal of Microelectromechanical Systems*, 2(1):33–43, 1993.

[6] C.H. Mastrangelo and C.H. Hsu. Mechanical stability and adhesion of microstructures under capillary forces – part ii: Experiments. *Journal of Microelectromechanical Systems*, 2(1):44–55, 1993.

[7] L.L. Howell. *Compliant Mechanisms.* John Wiley & Sons, Inc., New York, 2001.

Chapter 6

# DESIGN RULES

As mentioned earlier, surface micromachining involves the deposition and patterning of multiple thin-films. Patterning is typically accomplished through photolithography (chapter 2.1). Thus, the design of a surface-micromachined device consists of the geometry on the various masks used during the fabrication process.

Surface-micromachined devices are designed by creating a series of two-dimensional drawings. These drawings are then transferred to the masks used in photolithography. Design rules are simply constraints on those drawings, and represent limits on pattern transfer to the various thin-films. Some of the imperfections that occur when patterning thin-films are discussed in section 7.2.

Typical design rules include:

- Minimum width
- Minimum spacing
- Minimum enclosure
- Minimum cut-in
- Minimum cut-out

## 6.1 Minimum Geometry Rules

Minimum geometry rules cover two particular rules: minimum width and minimum spacing (figure 6.1). Minimum geometry rules exist to ensure that the features are large enough to be resolved during pattern transfer. For example, features smaller than the minimum feature size will not appear in the manufactured devices.

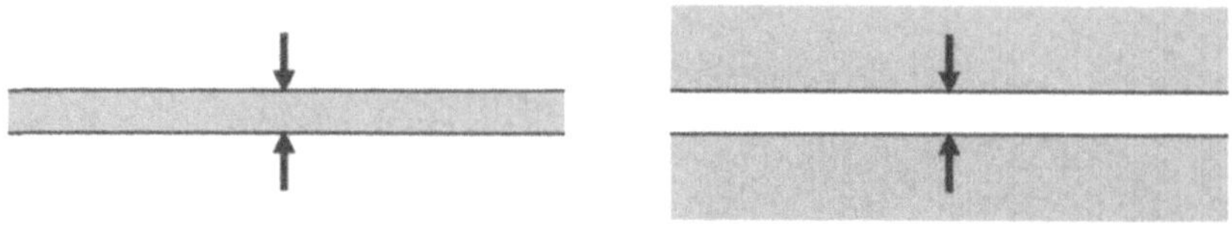

(a) Minimum Width (b) Minimum Spacing

*Figure 6.1.* Illustration of minimum geometry rules.

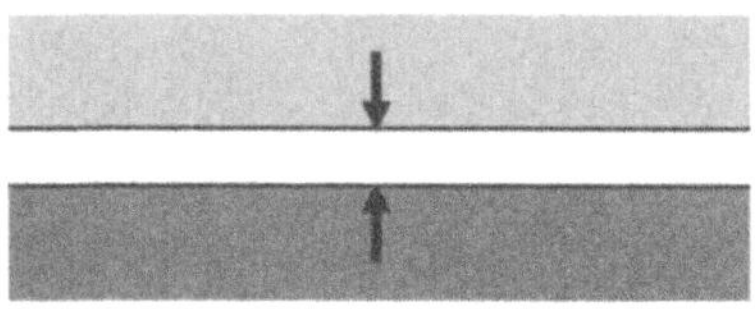

*Figure 6.2.* Illustration of minimum spacing rule between different mask layers.

Often, the nominal width and nominal spacing rules will also be specified. These represent safe values, and should be used instead of the minimum width and minimum spacing rules unless there exists a good design reason.

While minimum width rules always apply to a single layer, the minimum and nominal spacing rules may also apply between different layers (figure 6.2).

Designs that restrict themselves to meeting all of the nominal geometry rules[1] for width should return from fabrication as expected. Barring design errors, broken devices should be very rare.

Designs that restrict themselves to meeting the minimum geometry rules are on thinner ground, but in general can be expected to return from fabrication as expected. However, designs that only meet the minimum geometry rules do often fail. For example, in PolyMUMPs™ thin parts (for example, beams) that traverse regions with significant topography will fail. As described in the section on structural faults (section 7.1.2), thin beams that cross over large changes in height (>1 $\mu$m) are weak points that usually break. Providing sufficient structural support can partially compensate for the structural weakness as well as provide redundancy.

The design rules for minimum width and minimum spacing should always be followed. Designs that violate minimum geometry rules will have missing, undersized, oversized or fused features. In particular, most fabrication services will require that designs at least meet the minimum geometry rules. These rules are mandatory.

## 6.2 Maximum Geometry Rules

Although less common than minimum geometry rules, some microfabrication processes will also specify maximum dimensions as well.

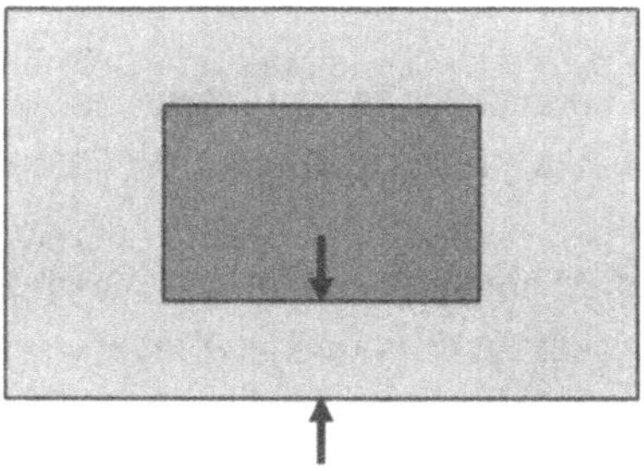

*Figure 6.3.* Illustration of minimum enclosure rules.

(a) Minimum Cut-in

(b) Minimum Cut-out

*Figure 6.4.* Illustration of minimum overlap rules.

For example, maximum width rules are common in surface micromachining. During the release step, it is important that the isotropic etchant can reach and remove all the sacrificial material in a reasonable time. Maximum width rules ensure that the etchant atoms or molecules do not need to travel very far. This ensures that the release step can be done in a fixed time.

## 6.3 Minimum Enclosure Rules

Where features on one mask surround, or enclose, features on another mask layer, enclosure rules specify a minimum width for this region.

Where minimum geometry rules ensure that the resolution of the photolithography process is sufficient for fabrication, the minimum enclosure rules (figure 6.3) ensure that the alignment between masks is sufficient. For example, enclosure rules for anchors ensure that the anchors are completely covered by polysilicon. When enclosure rules are not followed, anchors can be partially exposed to subsequent etching during polysilicon patterning.

The minimum enclosure rules may overstate the inter-mask alignment tolerance to also handle topography effects. For example, photolithography and pattern transfer is more difficult near changes in the wafer's height. Thus, minimum enclosure rules can push geometry boundaries away from these regions.

## 6.4 Minimum Overlap Rules

Where features from one mask cross the boundaries from features on another mask, overlap rules create minimum distances that features must extend past any boundaries that they cross (figure 6.4).

Overlap rules[2] exist to prevent extreme height changes, which will interfere with lithography, and to prevent unnecessary etching of underlying layers.

Despite the above, the minimum overlap rules are considerably more flexible. While they certainly should be followed wherever possible, there exist many reasons to violate these rules. In particular, the PolyMUMPs$^{TM}$ Design Handbook states that the minimum overlap rules are only advisory rules[3]. Users are free, at their own risk, to violate them.

## 6.5 Design Rule Checks (DRC)

Design Rule Checks are an excellent way to validate your design before submitting it for fabrication. Software that can automatically perform design rule checking can detect errors early. Unfortunately, DRC technology for micromachining is immature and does not accurately check many designs.

First, there are several types of errors that DRC do not find. For example, DRC will not detect cases of minimum width beams crossing height changes that may constitute a structural fault.

Second, there are several types of spurious errors that DRC may find. Most annoying is DRC software that cannot properly handle non-Manhattan[4] geometries. Where this is the case, curved edges can create thousands of spurious errors.

Also, in PolyMUMPs$^{TM}$, to prevent common beginner's errors, DRC rules often have checks on anchor and via layers. These rules will create errors wherever the nitride is breached or a double thickness structure is placed.

## Notes

1 Nominal geometry rules include nominal width and nominal spacing rules.

2 Overlap rules include enclosure, cut-in, and cut-out rules.

3 For example, the PolyMUMPs™ Design Handbook states that inter-level spacing rules are advisory. However, this statement is considerably weaker. Inter-level spacing rules should be treated the same intra-level spacing rules, and thus should be mandatory.

4 Manhattan geometries consist of rectangles aligned with the axes.

Chapter 7

# NON-IDEAL PROCESSES

## 7.1 Topography

In many types of thin-film deposition, film growth is conformal. The shape of a deposited layer follows the topography of the underlying wafer surface when deposition occurred. The result of any processing step therefore depends heavily on the preceding steps. Managing topography is an important part of the design process. Generally, design faults relating to topography can be classified into three categories: interference faults, structural faults, and behavioural faults.

While surface micromachined systems are typically built up, in some cases the steps that follow can affect deposited layers. An example of this is the unintended etching of deposited layers when protecting etch stops are removed.

### 7.1.1 Interference Faults

The first category of design faults consists of interference faults. During deposition, thin films conform closely to the topography of the previously deposited and patterned layers, and so are not necessarily planar. Often, the topography can trap a structure that was intended to move freely. Unless the preceding layers are designed to ensure the upper structural layers are flat where needed, the induced topography can have detrimental effects on device operation.

Figures 7.1 and 7.2 are two examples of devices with interference faults. In figure 7.1, the beam extending past the top and bottom of the image is suppose to move up and down. Notice how the beam is folded over the edges at the two locations where it first meets the perimeter of the square structure. The beam clearly cannot move.

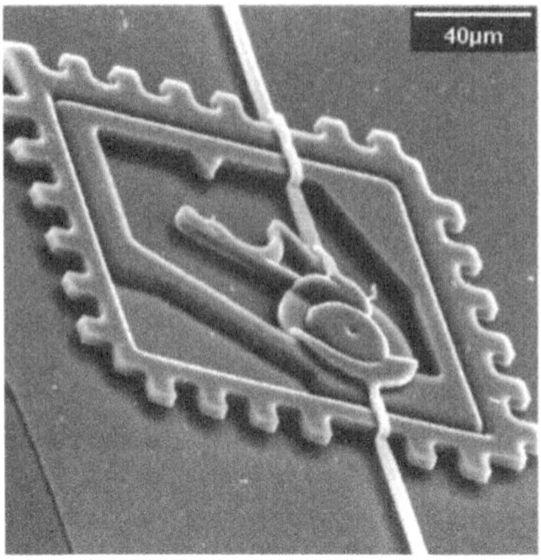


*Figure 7.1.* Part of bi-stable switch design with rod trapped by conformal deposition. Fabricated using PolyMUMPs™.

*Figure 7.2.* Gear design trapped by two rods due to conformal deposition. Fabricated using PolyMUMPs™.

*Figure 7.3.* Gear redesigned to avoid interference faults, but now showing structural faults. Fabricated using PolyMUMPs™.

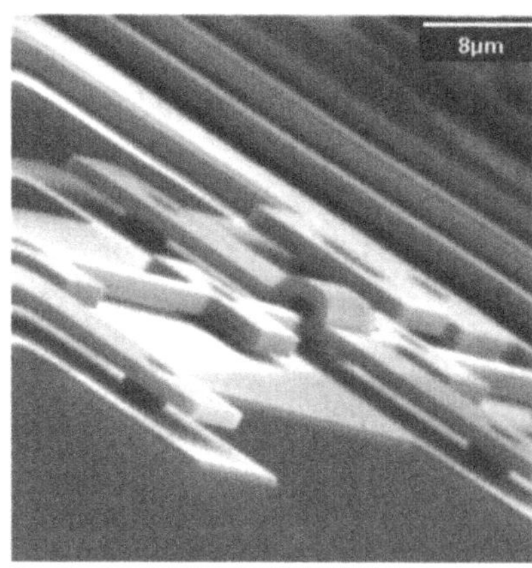


*Figure 7.4.* Staple interfering with lateral motion of a rod. Fabricated using PolyMUMPs™.

Figure 7.2 shows a similar problem. While the pin appears to be properly formed, the two connecting beams have conformed to the gear's teeth, preventing rotation. A redesign of this gear is shown in figure 7.3.

Figure 7.4 shows a different type of interference fault. Here, the staple exists to keep the beam close to the wafer surface. This type of support can be necessary to prevent damage during fabrication and release. However, the sta-

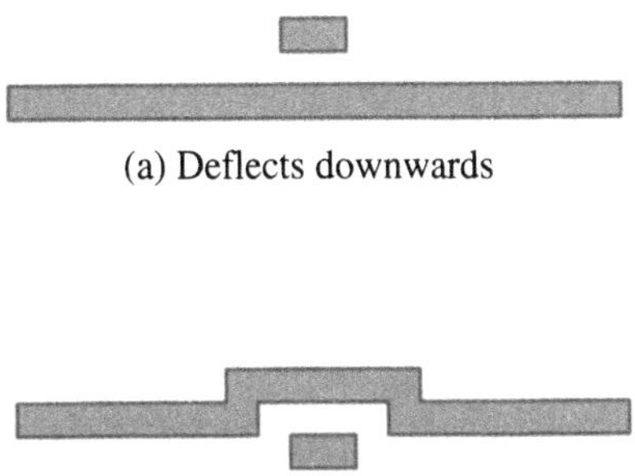


*Figure 7.5.* Cross-sections for vertical moving electro-thermal-compliant actuators.

*Figure 7.6.* SEM of a pair of vertical electro-thermal-compliant actuators. Fabricated using PolyMUMPs™.

ple has conformed over the rod and prevents any lateral motion. Although not discernable in this image, this rod required lateral motion to operate properly.

Some surface micromachining processes, for example SUMMiT™, may planarize the wafer after key steps. Certainly, the need for planarization increases with the number of structural layers. However, unless the wafer is planarized once for each structural layer, topography will still need to be managed.

### 7.1.2 Structural Faults

Structural faults, the second category of topography problems, occur when the topography creates a structural weakness in the fabricated device. Not only can the topography provide stress concentration points, beam thinning may occur due to variation in film thickness across steps. These problems are compounded since lithography is more difficult along height changes. At least in PolyMUMPs™ minimum thickness beams crossing significant steps are likely to break.

While the design shown in figure 7.3 has removed the interference faults shown in figure 7.2, the design now has structural faults. This is the cause of the break in the fourth connecting beam. Fortunately, this device may still work. Using more than one connecting beam has provided a measure of redundancy. Further, the use of multiple connections provides structural support, which reduces the chance that any individual beam will break.

*Figure 7.7.* Cross-section of pivot joint showing how topography can reduce tolerance.

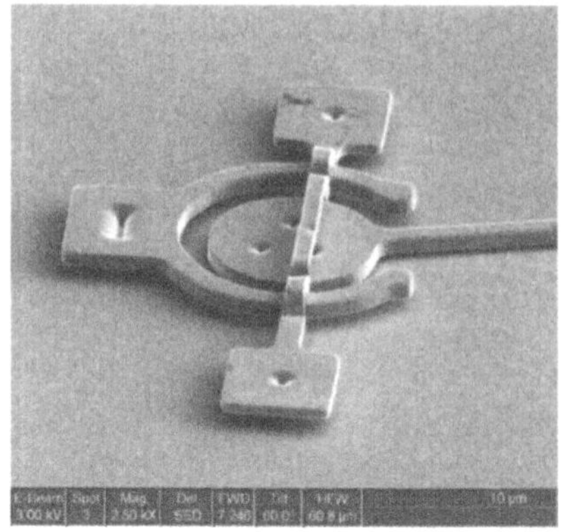

*Figure 7.8.* SEM of a revolute joint that uses topography to reduce tolerance. Fabricated using PolyMUMPs™.

## 7.1.3 Behavioural Faults

Behavioural faults make up the third category. Figure 7.6 shows two electrothermal actuators. The top actuator is supposed to move upwards when power is applied; the bottom actuator should move downwards.

Like the bottom actuator, the top actuator is composed of a wide arm and narrow arm. Differential heating due to an applied current causes differential thermal expansion (section 12.2). This was supposed to cause the arm to curve upwards. However, the wide arm has conformed and is no longer flat. The wide arm's bending stiffness is thus significantly higher, reducing any motion.

The bottom actuator, except for the fact that it is stopped by the wafer surface, works as expected. While the design is essentially the same, the reversed roles of the two beams have ensured that both arms of the actuator are flat.

Another good example of behavioural faults is mirrors placed over other structures. Mirrors are ideally flat. However, buried structures will lead to variations in the mirror surface.

## 7.1.4 Intentional Use

Topography can also be used be used to create beneficial changes in behaviour. For example, topography that causes change in the cross-section of a beam can be used to increase the beam's bending moment of inertia, and so stiffen the beam [1]. This can be used to make a beam more resilient to adhesion and non-uniform residual stresses.

The problematic topography shown in figure 7.6 could thus be advantageous in another situation.

(a) Example of a positive sidewall offset

(b) Example of a positive sidewall angle

*Figure 7.9.* Illustration of two types of lateral dimensional errors.

Another beneficial use for topography is overcoming minimum feature spacing rules of a process (section 6.1). If the following sacrificial layer is thinner than the minimum feature spacing, then designers can use topography to create interlocking parts that are closer together than allowed by minimum feature spacing. This can be used to create interlocking gears or sliding mechanisms [1].

For example, note how the top structural layer has filled in the interior space of the revolute joint shown in figures 7.7 and 7.8. The minimum feature spacing for this design was 2 $\mu$m, but the second structural layer has filled in these spaces and so the joint's lateral freedom is $\pm 0.75$ $\mu$m, instead of the $\pm 2$ $\mu$m it would have been otherwise.

## 7.2 Dimensional Tolerances

Just like all manufacturing processes, surface-micromachining does not reproduce parts with exactly the specified dimensions. Some types of devices show large sensitivity to manufacturing tolerances [2]. This difficulty is compounded by the large relative tolerances in surface-micromachining, where fabrication tolerances can be as large as 20% of the nominal width (section 6.1) are common [2]. As an example, test structures for determining material properties had errors in the extracted material property values as high as 100% [2].

Surface-micromachined devices are subject to a number of dimensional tolerances. The first, and simplest, tolerance is the film thickness. Depending on the method of deposition (section 2.3) and amount of in situ measurement, films can be grown with atomic thickness tolerances. However, tolerances of 10% are common.

While film thickness is controlled by the deposition method, lateral dimensions are controlled by the method of pattern transfer. This involves both photolithography (section 2.1) and etching (section 2.4). Three tolerances are considered here:

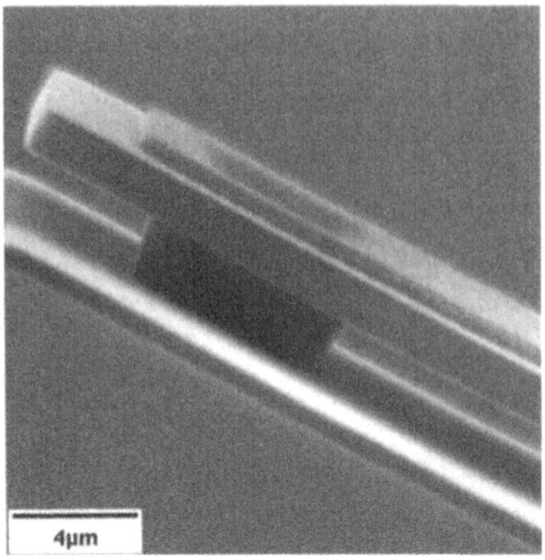


*Figure 7.10.* SEM of an anchor showing negative sidewall angle. This is expected from conformal deposition of a via with positive sidewall angle. Fabricated using PolyMUMPs™.

- The first tolerance is *magnification*. During photolithography, the pattern is transferred optically from the mask to the photoresist. The pattern is not necessarily projected into the photoresist at the same size. This may be done intentionally, to reduce geometry, or unintentionally. If the image magnification is unintentional, then this leads to an error. Magnification leads to relative errors, in that the size of the dimensional tolerance is proportional to the length in question.

- The second tolerance is the sidewall or process *offset* (figure 7.9). An offset tolerance is when the edge of a part is offset from its intended position [2]. An example cause of offset is an etching process that undercuts or overcuts the photoresist.

  Offset tolerances typically affect small dimensions more than large dimensions since it is an absolute tolerance. For example, consider the common case of a beam. If the beam's nominal width is $w_0$, and the offset is known to be $w_{off}$, then the beam's actual width can be found as follows. Notice that both edges of the beam are affects by the sidewall offset.

$$w = w_0 + 2w_{off} \tag{7.1}$$

- The third tolerance is *sidewall angle* (figure 7.9). Sidewalls that are perfectly perpendicular to the substrate surface is an idealization. In reality, the sidewalls can have arbitrary shapes[1]. However, it is often assumed that the sidewalls are flat, but at an angle with respect to the perpendicular[2] [2].

## 7.3 Thin-film Stresses

A basic understanding of the relationships between stresses and strains is required. The basic theory is fairly simple and can be obtained from almost any elementary structural mechanics text [3].

Unfortunately, the thin-films that comprise surface micromachined structures will often be stressed. These residual stresses will lead to shape changes when devices are released. Residual stresses are due to a large number of factors, not all of which are fully understood.

For example though, the easiest source of residual stress to understand is due to different coefficients of thermal expansion (CTE) amongst materials. If the films are considered to be unstressed at the deposition temperature, then the residual stresses are the result of cooling the wafer to room temperature. Since the materials have different CTEs, they will shrink by different amounts during cooling. When the wafer bulk is much thicker than the thin-films, it can be considered essentially rigid. Since the films are evenly attached to the wafer surface, the thin-films must therefore stretch (tensile stress) or shrink (compressive stress) to match the dimensional changes of the wafer.

Residual stresses perpendicular to the wafer surface are usually ignored in surface micromachining. Since structures are not constrained in this direction, their dimensions are free to move to their relaxed state. Because they are thin, the dimensional changes are usually negligible. Because they are relaxed, the stress is zero.

However, stresses parallel to the wafer surface can have significant consequences for devices. Stress effects, such as buckling or curling, occur regularly and many structures are large enough that dimensional changes are significant. The effects of residual stresses are typically understood in terms of the uniform stresses and the non-uniform stresses[3].

Since the wafer bulk is essentially rigid, thin-film stresses are typically measured by comparing the film's strain against the bulk of the substrate.

### 7.3.1 Uniform Residual Stress

The uniform stress is the average stress through the thickness of the film. While it may be unsafe to consider it constant across the entire wafer, it should be vary slowly. For singly supported structures, one should expect a dimensional change as the structure relaxes to a non-stressed state (figure 7.11). Doubly supported structures can be more reliable, in that their length will remain fixed. However, over a critical compressive stress they will buckle.

Fortunately, the uniform stresses of the various films should be somewhat controlled. For example, the PolyMUMPs™ Design Handbook outlines the

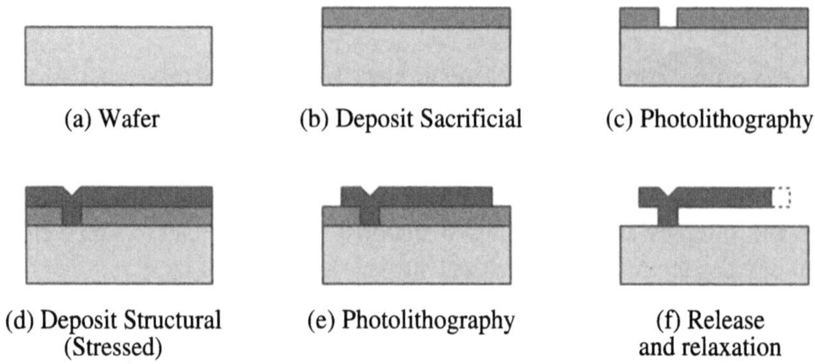


*Figure 7.11.* Fabrication of a cantilever using a thin-film with tensile stress. Note that cantilever shrinks after release.

expected range of the residual uniform stress in the structural layers of the PolyMUMPs process. The residual stresses are measured for each fabrication run and publicly posted.

Since cantilevers are free on one end, they are also free to relax. Thus, any uniform thin-film stresses in the cantilever will relax. This will lead to a change in dimensions (figure 7.11).

The dimensions of a cantilever as described by the mask describe that cantilever during pattern transfer. When the cantilever is released, it will relax and undergo dimensional changes. Most of these can be ignored. However, the length may change enough to cause problems.

For a singly supported structure, the dimensional changes that result from a given stress can fortunately be easily calculated. The change in length can be determined by using the following formula [3]:

$$\epsilon = \frac{\sigma}{E} = \frac{\Delta}{L} \tag{7.2}$$

In this formula, $\epsilon$ is the thin-film stress[4], $\sigma$ is the thin-film strain, $E$ is Young's modulus, $\Delta$ is the change in length of the cantilever, and $L$ is the unsupported length. Since the material properties ($\epsilon$, $E$) are known process parameters, and $L$ is a design parameter, the expected change in the cantilever's length can be determined. Please note that in equation 7.2, since the nominal dimensions corresponds to the stressed state. Released structures will have a length change opposite to that calculated using equation 7.2.

For doubly supported structures, the key parameter that needs to be considered is the critical buckling stress. If the film's stress is compressive, and exceeds the critical strain, the structure will buckle (figure 7.12). Pleaser refer to the section on buckling in section 5.4.1.

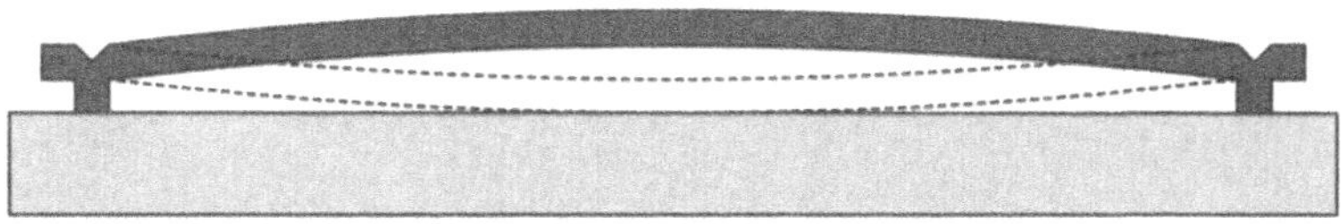

*Figure 7.12.* A doubly supported structured (Euler column) showing buckling.

For doubly supported structure, the two supports must provide sufficient force to either provide the required dimensional change or the buckling strain. Unlike in the singly supported structure, the structural strength of the supports in doubly supported structures may be of concern.

Beyond any dimensional changes, residual stresses will continue to affect the dynamic behaviour of any devices. Through effects such as strain stiffening, residual stresses will change the mode frequencies.

For more complex structures, numerical modeling may be required. The software for structural simulations is well developed and the simulations are relatively easy to setup and run.

The fabrication process does not oversize or undersize design components to account for these dimensional changes. Designs need to be robust enough to work with the range of residual stresses that may be present in the end result. Residual stresses are usually in the range of tens of MPa [4], but can be much higher.

### 7.3.2 Non-uniform Residual Stresses

Changes in the substrate temperature during film deposition, changes in the stoichiometry, changes in material density before annealing, and other effects, will lead to changes in the film's strain throughout its thickness. This is equivalent to an applied bending moment on the layer. When the structures relax upon release, they will have a curvature.

The net bending moment can be calculated using equation 7.3.

$$M = \int_0^t \sigma(y)\, w \left(y - \frac{t}{2}\right) dy \qquad (7.3)$$

Above, $M$ is the bending moment, $\sigma$ is the film stress at height $y$, $w$ is the width of the structure, and $t$ is the film's thickness. While valid, equation 7.3 may not be of much direct use as knowledge of the stress distribution, $\sigma(y)$, will be difficult to obtain. However, experimental methods, including test structures, may be able to provide an empirical value for the net bending moment.

If the bending moment is known, the deflection of a singly supported structure can be predicted relatively easily. As the structure will have a uniform bending moment throughout its length, the relatively well-known beam deflection formula can be used.

$$\delta(x) = \frac{Mx^2}{2EI} \tag{7.4a}$$

$$\delta(x) = \frac{M\left(1-\nu^2\right)x^2}{2EI} \tag{7.4b}$$

Where $\delta$ is the beam's deflection, $M$ is the bending moment, $x$ is the distance from the structure's support, $E$ is the modulus of elasticity, $I$ is the moment of inertia, and $\nu$ is Poisson's ratio. For thin beams, equation 7.4a should be used. However, if the beam is considerably wider than the film's thickness, than the biaxial modulus of elasticity should be used, which leads to equation 7.4b.

If the approximation that the stress varies linearly through the film is appropriate, then the stress can be written as:

$$\sigma(y) = \left(\frac{\sigma_1+\sigma_2}{2}\right) + \left(\frac{\sigma_1-\sigma_2}{t}\right) y \tag{7.5}$$

Above, the stress, as a function of distance, is determined based on the $\sigma_1$ and $\sigma_2$, the stress at the top and bottom of the film, respectively [5]. The variable $t$ is the film's thickness. Using this equation, and equation 7.3, one can determine the bending moment. For a beam with a rectangular cross section, this results in the following equation [5]:

$$M = \frac{(\sigma_1-\sigma_2)wt^2}{12} \tag{7.6}$$

In equation 7.6, the bending moment is calculated based non-uniform stress, the beam's width, $w$, and the beam's thickness, $t$.

Non-uniform stresses are more troublesome than uniform stresses. Non-uniform stresses are both more difficult to handle theoretically and more difficult to measure. For example, while the residual uniform stresses are limited in PolyMUMPs$^{TM}$ (specified in the Design Handbook) and are measured for each fabrication run, no equivalent controls are placed on non-uniform stresses.

Gradient stresses are always present, and can be severe enough to impede normal device operation. For example, figure 7.13 illustrates a beam that was supposed to be able to move sideways. The beam has, however, curved up into the space created by its topography. As structures, including beams, reach lengths of 100 $\mu$m, curvature becomes significant.

More commonly, non-uniform stresses affect gears and mirrors, where it causes bowing. While the radius of curvature will be less in these essentially

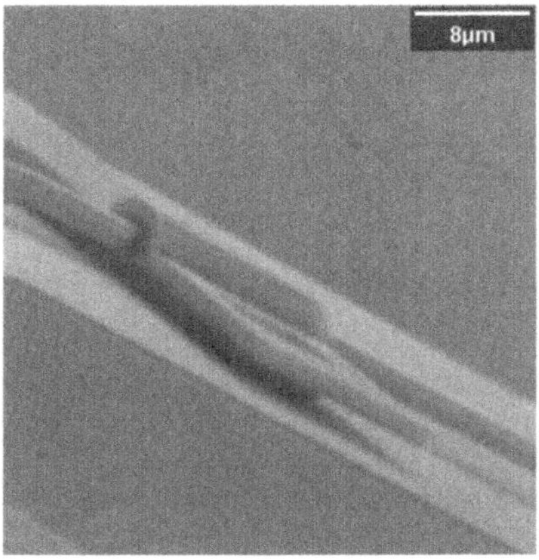


*Figure 7.13.* SEM of a beam, which, due to non-uniform stresses, has curled up, preventing sideways motions. Fabricated using PolyMUMPs™.

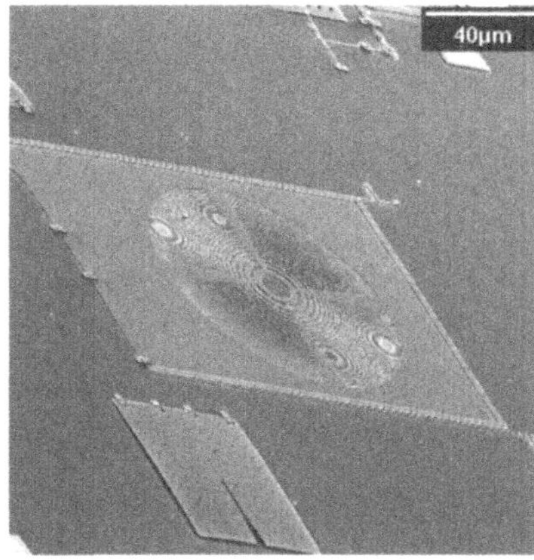


*Figure 7.14.* SEM of a Fresnel-zone plate, which is relatively flat compared to its vertical support. Fabricated using PolyMUMPs™.

2D structures than the equivalent 1D rods, the effects can be more troublesome. For gears, excessive bowing will cause teeth to slip. For mirrors, bowing will preclude their use in most optical applications.

Figure 7.13 shows a beam with convex curvature. This figure is related to figure 7.4 as the staple is causing an interference fault. However, the problem is compounded in this case by the beam's curvature, which pushes the beam up into the space created by the conformal deposition of the second structural layer. Even if the second layer were flat, the curvature would probably push the beam against the staple. The likely result would be adhesion and consequently the immobilization of the beam.

Where flatness is required, some steps can be taken to reduce curvature. Notice how in figure 7.14 the Fresnel-zone plate is considerable flatter than its vertical support. Since the plate is significantly larger, one would expect a larger deflection. However, the designers have reduced the curvature by stiffening the plate. The second structural layer was used around the plate's edges to increase its thickness.

## Notes

1 The sidewalls of structures etched using DRIE tend to have scalloped or textured surfaces.

2 This approximation works well because the sidewall angles of RIE and DRIE processes are near zero.

3 Non-uniform stresses are also known as gradient stresses.

4 By convention, positive values for the stress indicate tension. Negative values indicate compression.

## References

[1] J.H. Comtois and V.M. Bright. Design techniques for surface-micromachining mems processes. *Proceedings of the SPIE*, 2639:211–222, 1995.

[2] R.K. Gupta. Electronically probed measurements of mems geometries. *Journal of Microelectromechanical Systems*, 9(3):380–389, 2000.

[3] S.P. Timoshenko and J.M. Gere. *Mechanics of Materials*. Van Nostrand Reinhold, New York, 1972.

[4] Y.B. Gianchandani. Bent–beam strain sensors. *Journal of Microelectromechanical Systems*, 5(1):52–58, 1996.

[5] A.K. Chinthakindi, D. Bhusari, B.P. Dusch, J. Musolf, B.A. Willemsen, E. Prophet, M. Roberson, and P.A. Kohl. Electrostatic actuators with intrinsic stress gradient - i. materials and structures. *Journal of the Electrochemical Society*, 149(8):H139–H145, 2002.

Chapter 8

# DESIGN TECHNIQUES

## 8.1 Ground Planes

The existence of the first structural layer in the PolyMUMPs$^{TM}$ process is specifically for making ground planes, which are necessary to electrically shield devices from the wafer [1–3]. Any two conducting bodies at different potentials will experience an attractive force; this includes surface micromachined devices and the substrate. If the attractive force is strong enough, the device will be pulled down to the substrate surface. Even if the device does not adhere, significant friction will be present.

An electrical contact can be created near the device by breaching the nitride. However, breaching the nitride is not a guaranteed process and so presents reliability issues. Further, trapped charges may exist at the silicon to silicon-nitride interface. These charges will create electrical fields, which, through induced charges on the device, again lead to an attractive force. Ensuring that the substrate and devices are at the same voltage will not necessarily eliminate all electrical fields. Hence, ground planes are needed.

Since the grounding plane is conducting, it cannot be used indiscriminately; otherwise all electrical connections on the chip would be shorted. Since the main concern with substrate fields is the resulting friction, only moving devices require a ground plane.

The electrical potential of ground planes should not be left floating. Unless their voltage is fixed, they too can become a source of electric fields and attract devices. Many devices will actually have a voltage distribution, in which case one simply aims to reduce the attraction to a reasonable minimum.

Figure 8.1 shows two electro-thermal actuators, one that has a ground plane and one without. The actuator without the ground plane is no longer freestanding. Although not in view, the ground plane is connected only to the pin nearest

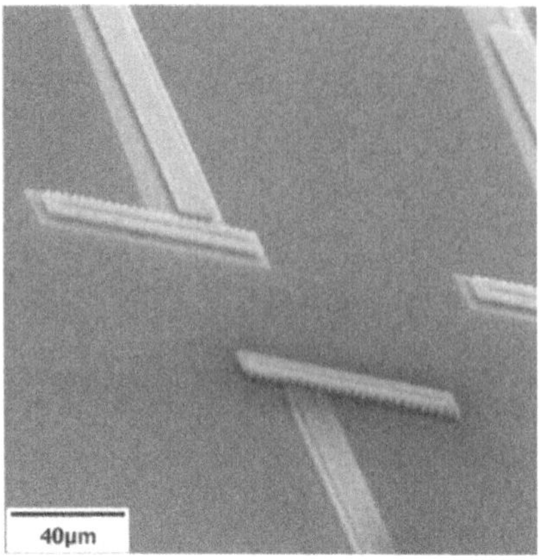


*Figure 8.1.* SEM of two electro-thermal actuators. The top actuator has an ground plane, while the bottom actuator does not. Fabricated using PolyMUMPs™.

the wide arm. This design is simple and sufficiently limits the attraction. However, this design does not completely eliminate the attraction, and may not be sufficient for larger dimension.

## 8.2 Double Thickness Parts

In general, contact surfaces should be avoided in surface micro-machined devices because of the large relative strength of surface forces, such as friction. However, where contact surface are needed, double-thickness parts will often be needed. Because surface micromachined devices are so thin, they will have significant vertical flexibility. Further, there may be bowing due non-uniform stresses. To ensure contact points stay in contact, double-thickness parts can be invaluable.

At least in PolyMUMPs™ the main difficulty with double-thickness parts is the region of P1P2VIA that must surround them[1]. The region can make the transition from double thickness region to single thickness regions tricky. Improper design easily leads to additional design rule violations and structural faults. Dealing with double-thickness parts can thus be a little awkward. While double-thickness parts are not guaranteed, they are fabricated reliably.

Figure 8.2 shows several surface micro-machined gears. The gear on the right has double thickness teeth, while the one on the left has single thickness teeth. Despite the relatively small size of the gears presented in figure 8.2, the single-thickness teeth exhibit some offset. While the single-thickness teeth are meshing properly in this device, larger gears or a larger non-uniform gradient stress will lead to slippage. The double-thickness gears are more firmly in contact.

A close-up of the transition between single-thickness and double-thickness teeth is shown in figure 8.3.

*Figure 8.2.* SOM of single thickness and double-thickness gears combined. Fabricated using PolyMUMPs™.

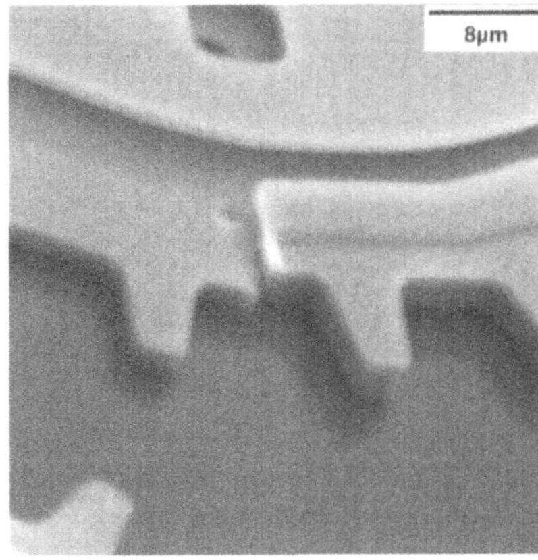


*Figure 8.3.* Close up SEM of transition region between single and double thickness teeth of a gear. Fabricated using PolyMUMPs™.

## 8.3 Conducting Paths

Providing electrical signals is an important part of MEMS design. In the traditional polysilicon processes, the polysilicon layers are heavily doped and have relatively low resistivities. However, when it comes to conductivity, the polysilicon still cannot compete with metal. Thus, it is important to use metal in conducting paths whenever possible to lower path resistances.

Further, the layout for a pad in PolyMUMPs™ is fairly specific. Many of the same issues that affect this supposedly simple layout also affect the design of conducting lines. Improperly designed conducting lines will break during the bonding process. With the exception of POLY0, all of the same layers should be present in conducting lines.

The widths of layers in the conducting line outlined here for PolyMUMPs™ should be increased beyond what is required to meet design rules[2]. With the minimum widths, the topography at the center of the conducting line is such that the metal deposition is affected, increasing the line's resistance. To ensure a low resistance, all widths should be increased by 2 $\mu$m, as this ensures that the center of the conducting line is flat. The suggested widths for each layer are listed in the table 8.5.

The conducting line outlined here has a current limit of several 100 mA at room temperature. Above this limit, the metal layer is damaged so that it can no longer carry any current. When this happens, the resistance of the conducting path increases significantly as the current must be carried by the polysilicon.

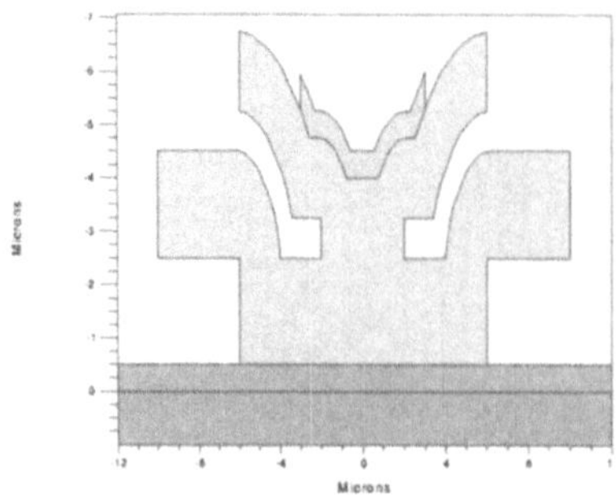

*Figure 8.4.* Structure of a conducting fabricated using MUMPs$^{TM}$ line after release.

| Layer | Width |
|---|---|
| POLY1 | 20 |
| ANCHOR1 | 12 |
| POLY2 | 12 |
| P1P2VIA | 4 |
| METAL | 6 |

*Figure 8.5.* Recommend layer widths for MUMPs conducting line.

While unlikely for micro-devices, if a larger current limit is required, one needs only to increase the width of the conducting lines.

Sometimes it is necessary to use only polysilicon as a conducting path. This occurs when paths need to cross under or between other layers. When necessary, the resistance can be calculated from the geometry and the published resistivity data. Obviously, wider paths are less resistive. Also, corners and contact methods can affect the resistance of the path.

## 8.4 Tethers

When designing structures that are loose, for example structures that are to be raised, the components will move around during release. This can damage the device itself as well as nearby devices. To prevent this, these devices should be tethered to the wafer surface. These tethers are broken just before the device is to be assembled or used for the first time.

Although many techniques are possible, there exist three main methods of breaking the tethers. The most common requires operators to break the tethers by using a microprober station. The probe needles are dragged across the surface of the wafer and across the tethers. Alternatively, the probe needle is dropped on the tether's anchor, crushing it.

When breaking tethers in this manner, a buffer zone is required around the tethers. Unless sufficient space exists around the tethers, the probe needles will damageDamage the surrounding structures. The tether should be at least 100 $\mu$m long. There should be at least 50 $\mu$m, and preferably 100 $\mu$m, of surrounding space.

This technique is relatively easy, and the equipment exists in most clean and semi-clean facilities. While popular, this technique has several flaws. First, this technique obviously requires operator intervention. Second, this technique can create a significant amount of debris. It is virtually impossible to remove this debris, which may later contaminate other devices.

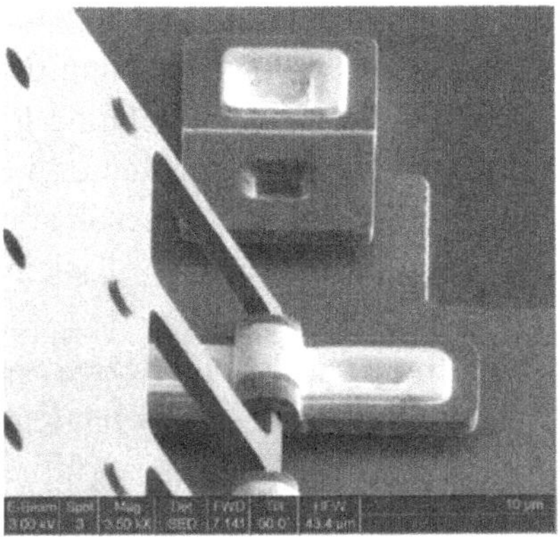

*Figure 8.6.* SEM of the underside of a raised plate. Note the presence of dimples. Fabricated using PolyMUMPs™.

*Figure 8.7.* SEM of a part of a suspension showing a dimple. Fabricated using SUMMiT™. Courtesy Sandia National Laboratories, SUMMiT™ Technologies, www.mems.sandia.gov.

Another technique is laser ablation. While this avoids the debris, the necessary equipment is not as readily available.

A third technique is to design the tethers like fuses. A large current is passed through the tethers, which heats the tethers until they break. While this technique is easily automated and leaves little debris, it may not always be ideal. Since tethers are for freely moving structures, ensuring a good electrical connection to the device may not be possible.

## 8.5 Dimples

Dimples[3] are small bumps on the underside of the first structural layer. For example, in PolyMUMPs™, a short wet etch is used to isotropically etch small 0.75 $\mu$m cavities in the first sacrificial layer [4]. The underside of the first structural layer will then have bumps, as it will conformally fill the holes. Because of the conformal deposition process, some measure of the dimple will also exist on the second structural layer.

Although there exists many uses for dimples, their main purpose is to prevent stiction. Whether or not a structure can be pinned to the surface through adhesion is determined by the elastic energy and surface energy of the structure. Elastic deformation of a structure stores energy, just like any conventional spring. Deforming a structure raises its potential energy. However, if the structure can be deformed so that it touches the wafer surface, and the surface energy lost is greater than the necessary elastic energy, then it is energetically favourable for the structure to remain collapsed.

For processes that undergo normal wet release and drying, ensuring that that the collapsed state is energetically unfavourable is essential. During the drying process, capillary forces will draw structure down. While the drying process is itself highly dependent on the liquid's surface tension, the result after complete drying depends only on the adhesive forces and structural elasticity. It is therefore essential that the collapsed state is unstable.

For devices that use the super-critical drying method, capillary forces no longer play a role. It is at least theoretically possible to successfully fabricate devices that would otherwise be pinned to the wafer surface. However, for reliability reasons, it is still better to ensure that the pinned state is unfavourable.

For a number of standard structures, the peeling bound can be calculated. Structures will not adhere for peel numbers greater than one. For a doubly supported beam, the peel number is:

$$N_P = \frac{128}{5} \frac{Eh^2t^3}{\gamma_s l^4} \tag{8.1}$$

Above, $N_P$ is the peel number, $E$ is the modulus of elasticity, $h$ is height of the beam from the wafer surface, $t$ is the beam's thickness, $\gamma_s$ is the interfacial adhesion energy per unit area, and $l$ is the distance between the two supports. If the beam's height is much larger than its thickness, then a different formula, which can be found in [5, 6], should be used. The interfacial adhesion energy is very high for crystals, and is on the order of 500 to 2000 mJ/m$^2$.

Other structures will have different formula for their peel numbers. For details, please refer to [5, 6]. In particular, the required dimple spacing for plate will be different from the dimple spacing required for a simple beam.

For the PolyMUMPs™ process, equation 8.1 can be used to calculate the maximum dimple spacing. Since the boundary condition is $N_P = 1$, $l$ can be at most 334 $\mu$m. However, equation 8.1 implies a completely fixed support, whereas a dimple only provides partial support. For this reason, and to add a margin of safety, dimples are typically spaced at 30 $\mu$m.

While it is important to ensure that enough dimples are present to prevent adhesion, their number should be minimized. Dimples in contact with the wafer surface may adhere, and so dimples can also be a significant source of friction.

## 8.6 Process Variations

Process variations can play on important role in the design of a system (section 7.2). For example, the average thickness of the different layers can vary by more than $\pm 10\%$ from run to run in PolyMUMPs™. Also important is lateral dimensional control, where tolerances may be as large as ±20% of the nominal feature size [7]. There will also be variations across a die. Static and quasi-static structures can usually handle process variation by including a safety factor. Unfortunately, dynamic structures can be very sensitive to process variations [7].

Designs for large structures will also have to consider the effects of residual stresses. In particular, long beams are subject to buckling when compressively stressed. Even if buckling does not occur, stress-stiffening can change the mechanical behaviour of structures.

## Notes

1 To create a double thickness part, the sacrificial layer between the two structural layers is removed. When the top structural layer is patterned, the pattern is also transferred to the bottom structural layer.

2 To include all of the layers in the conducting path and meet the design rules, the required width is 18 $\mu$m.

3 A dimple may also be considered a specialized type of boss [8].

## References

[1] Y.-C. Tai, L.-S. Fan, and R.S. Muller. Ic-processed micro-motors: design, technology and testing. In *Proceedings, An Investigation of Micro Structures, Sensors, Actuators, Machines and Robots*, pages 1–6. IEEE, 1989.

[2] M. Mehregany, P. Nagarkar, S.D. Senturia, and J.H. Lang. Operation of microfabricated harmonic and ordinary side-drive motors. In *Proceedings fo the IEEE Micro Electro Mechechanical Systems Workshop*, pages 1–8. SPIE, 1990.

[3] R. Legtenberg, A.W. Groeneveld, and M. Elwenspoek. Comb-drive actuators for large displacements. *Journal of Micromechanics and Microengineering*, 6(3):320–329, 1996.

[4] D. Koester, A. Cowen, R. Mahadevan, and B. Hardy. *PolyMUMPs Design Handbook*. MEMSCAP, http://www.memsrus.com/mumps.pdf, 2002.

[5] C.H. Mastrangelo and C.H. Hsu. Mechanical stability and adhesion of microstructures under capillary forces – part i: Basic theory. *Journal of Microelectromechanical Systems*, 2(1):33–43, 1993.

[6] C.H. Mastrangelo. Adhesion-related failure mechanisms in micromechanical devices. *Tribology Letters*, 3(3):223–238, 1997.

[7] R.K. Gupta. Electronically probed measurements of mems geometries. *Journal of Microelectromechanical Systems*, 9(3):380–389, 2000.

[8] J.H. Comtois and V.M. Bright. Design techniques for surface-micromachining mems processes. *Proceedings of the SPIE*, 2639:211–222, 1995.

PART III

# DEVICES

# Chapter 9

# ELECTRICAL COMPONENTS

The goal of surface-micromachining is miniature mechanical systems, so why are electrical components of interest?

First, surface-micromachined mechanical devices will usually need to interact with electrical signals. The miniature mechanical components will need to be turned on and off, and a macroscopic on-off switch is not feasible. In addition to control and logic, many devices will also obtain their energy by consuming electricity.

Second, electrical components created using surface-micromachining can be easily fabricated on the same chip as the mechanical devices, allowing tighter integration between electrical and mechanical components.

Third, many electrical circuits presently use components which are difficult to manufacture using conventional microelectronics fabrication techniques. For example, inductors and variable capacitors are typically discrete components [1, 2]. These components are therefore located off-chip, and so consume a significant portion of a system's total volume[1]. Surface micromachining can be used to miniaturize these discrete components.

## 9.1 Bonding Pads

Bonding pads are relatively large structures used to create off-chip electrical connections (section 2.9). They are the interface between on-chip and off-chip electrical signals. In microelectronics, significant effort is invested in their design[2]. However, their design is relatively simple in surface micromachining.

Bonding pads should have a minimum resistance, but also must be able to withstand the mechanical forces that occur during bonding.

*Figure 9.1.* SEM of the corner of a bonding pad fabricated using PolyMUMPs™.

### 9.1.1 Exclusion Zone

During the bonding process, significant stresses are created and supported by the substrate. These stresses are not confined to the bonding pad itself, and are known to cause delamination and cracking of nearby structures. A region surrounding the bonding pad where these negative effects occur is known as the exclusion zone. Ideally, no structures other than the bonding pad are located within the exclusion zone.

Obviously, electrical contacts are necessary. Some types of structures must therefore exist in the exclusion zone. The connecting structures must be designed to withstand the stresses creating during bonding. However, guidance can be taken from the recommended bonding pad design.

For example, the recommended bonding pad for PolyMUMPs™ [3], includes all of the structural layers plus the metal layer (figure 9.1). This structure should be repeated in the conducting lines used to bring signal from the bonding pad to other on-chip components.

### 9.1.2 Parasitic Resistance

Fresh bonds can have parasitic resistances that are on the order of $10^{-2}$ $\Omega$ [4, 5], which is negligible.

However, the resistance will increase over time. The resistance of bonds annealed for 100 hours in $N_2$ and at 300°C increased to several Ohms [4]. More aggressive gases will accelerate the degradation of the bonding wire.

### 9.1.3 Parasitic Capacitance

If the substrate is conducting, then the bonding pad and the substrate will form a parallel-plate capacitor. By design, bonding pads are relatively large flat plates, and so this parasitic capacitance can also be relatively large. For insulating substrates, this parasitic capacitance can be neglected.

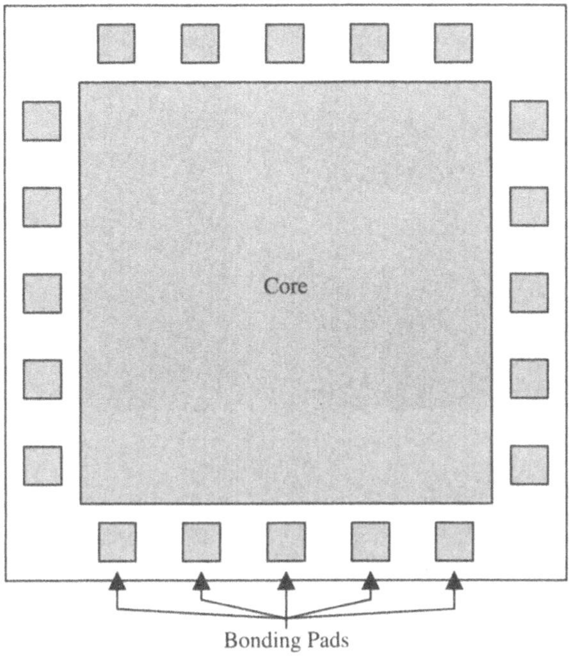


*Figure 9.2.* Illustration of a layout for a chip. Bonding pads are located around the perimeter while the functionality is located in the core.

Bonding pads are capacitively connected to the substrate through the isolation layer (section 3.3.3). If the substrate is conducting, which is common, then the substrate can form a common electrical node. Electrical signals applied to the pad structure may couple to other nearby structures, including other pads.

In addition to possible cross-talk, parasitic capacitance and resistive losses in the substrate will place an upper limit on the frequency of signal that can pass through the bonding pad [6]. The frequency limit can be increased by: (1) using insulating substrates, (2) removing the substrate from under the bonding pad, (3) using thick isolation layers, and (4) multilevel or thicker metallization.

The substrate capacitance can be determined by considering the substrate and pad as two plates (section 9.3) with the isolation layer forming the interplate dielectric. However, this forms a lower bound for the parasitic capacitance, as there are fringing fields, not only from the sides, but the top of the bonding pad. However, bonding pads are typically quite large, and fringing fields should be negligible.

*Figure 9.3.* SEM of a polysilicon resistor connecting two bonding pads fabricated using PolyMUMPs$^{TM}$. The resistor is 2 $\mu$m wide and 0.5 $\mu$m thick. Fabricated using PolyMUMPs$^{TM}$.

*Figure 9.4.* SEM of a polysilicon resistor with a serpentine path to save area. Fabricated using PolyMUMPs$^{TM}$.

### 9.1.4 Arrangement

Ideally, the bonding pads should be arranged around the outside edge of the chip (figure 9.2). This simplifies wire bonding (section 2.9). Ideally, the bonding pads should line up with the corresponding pads on the chip package.

Bonding pads can be placed in the chip's interior, but wire bonding will be more difficult. However, if electrical connections are made using a microprober station, then pad location is not critical. This is useful during development and testing, when devices are not necessarily packaged.

For commercial products, the bonding will usually done by automated equipment. This will typically impose further constraints on the pads, such as minimum size, minimum spacing, and a maximum distance from the edge of the chip.

## 9.2 Resistors

In a surface micromachining process, resistors can be made by creating bars of any conducting or semi-conducting material. However, designers should prefer layers with low conductivities, small thicknesses, and small minimum feature sizes. This will allow resistors to occupy a minimum amount of chip area (equation 9.1).

The resistance of a single rectangular line can be determined as follows:

$$R = \frac{\rho L}{wt} \tag{9.1}$$

In equation 9.1, $R$ is the resistance, $\rho$ is the layer's resistivity, $L$ is the bar's length, $w$ is the bar's width, and $t$ is the layer's thickness. In a surface-micromachining process, layer thicknesses will be constant, so it is common to combine the resistivity and layer thickness into a combined parameter. We thus have equation 9.2, where $R$ depends on the sheet resistance, $R_\Box$.

$$R = R_\Box \frac{L}{W} \tag{9.2}$$

$$R_\Box = \frac{\rho}{t} \tag{9.3}$$

If the resistance needs to be large, it often helps to "snake" the resistor (figure 9.4). For resistors with arbitrary paths, it is often convenient to calculate the resistance based on the sheet resistance. The resistance can be found by breaking the path up into squares, and then counting how many squares the current must pass through between the terminals.

At high frequencies, additional phenomena begin to affect a resistor's performance, in particular skin depth [7].

### 9.2.1 Sidewall Doping

When determining the resistance, it is important to check for sidewall doping. If the structural material is a semiconductor, it will be doped (section 2.5) to become electrically conductive. Depending on the process used to dope the semiconductor, the resistivity of the material may or may not be uniform.

In particular, the structural material can be doped while it is being deposited. This is called *in situ* doping, and should lead to a material that is initially uniform. However, even in this case, high temperature processing after the deposition can lead to changes through diffusion (section 2.5.1). Dopants may still leave or enter the material, changing dopant concentrations near the material's surfaces.

Also, it is not always possible to dope the material during deposition. Doping will then be performed after the material is deposited. Non-uniform doping should be expected in this case. For example, if the dopant atoms are introduced through diffusion, then the dopant atoms migrate through the surface to the material's interior under high temperature. Regions with more exposed surfaces, or closer to the surface, will have higher dopant concentrations, and therefore different conductivities.

Sidewall doping will affect the accuracy of equation 9.1. Instead, one should use the following equation:

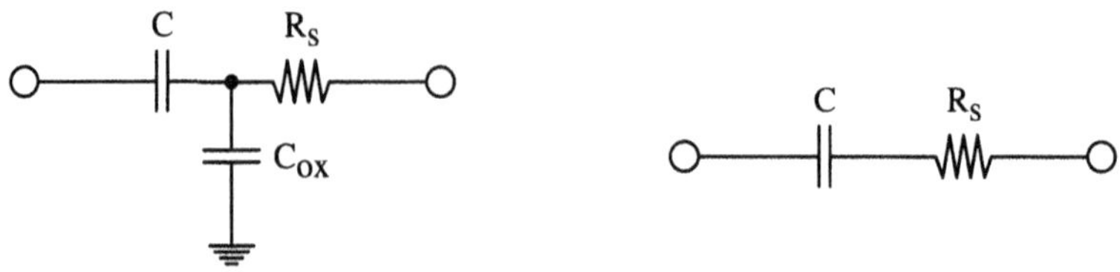


(a) Capacitor with substrate parasitics (b) Capacitor without substrate parasitics

*Figure 9.5.* Electrical models for non-ideal capacitors [6].

$$R = \frac{L}{t} \frac{1}{\int \sigma(x,y) \mathrm{d}x \mathrm{d}y} \tag{9.4}$$

In equation 9.4, the resistance, $R$, is computed from the resistor's length, $L$, thickness, $t$, and conductivity, $\sigma(x)$. The conductivity is a function of position across the resistor, $x$.

### 9.2.2 Parasitics

Surface-micromachined resistors will have both capacitive and inductive parasitics. Unfortunately, accurately determining these parasitics is difficult [7]. This difficulty stems, in part, from the reality that these parasitics are distributed.

A simple approach to estimating the parasitic capacitance is to simply treat the wires as plates. Wherever wires, or resistors, lie over the substrate, then then the area of the wire can be used to estimate the parasitic capacitance using the formula for parallel-plate capacitors. Similarly, wherever wires overlap, the parasitic capacitance between the two wires can be estimated.

## 9.3 Capacitors

At least to first approximation, the capacitance of the surface-micromachined capacitors can be determined using the formula for parallel plate capacitors:

$$C = \frac{\epsilon A}{d} \tag{9.5a}$$

$$C = \frac{\epsilon_r \epsilon_0 w l}{d} \tag{9.5b}$$

In the above equations, $C$ is the capacitance, $\epsilon$ is the permittivity of the dielectric, $\epsilon_0$ is the permittivity of free space, $\epsilon_r$ is the relative permittivity of the material, $A$ is the area of the capacitor, $w$ is the capacitor width, $l$ is the capacitor height, and $d$ is the separation distance between the plates.

(a) Parallel-plate Capacitor

(a) Wedge Capacitor

*Figure 9.6.* Illustration of the capacitors available in surface-micromachining.

The above formula is limited in that it does not account for fringe effects. In general, the effects of fringing for parallel plate capacitors in surface-micromachining can be neglected as long as both the width and length of the capacitor are significantly larger then the plate separation, a condition that is easily achieved. However, the presence of etch holes contributes additional fringing that will always be present as their presence increase along with the area of the capacitor.

If the fringing fields are important, they can be approximated by treating the edges of the capacitor as a wedge capacitor [8]. The capacitance of a wedge capacitor is:

$$C = \frac{\epsilon w \ln\left(L/l\right)}{\beta} \tag{9.6}$$

In equation 9.6, $w$ is the width of the capacitor's plates, $L$ is the length of the plates, $2l$ is the separation of the plates, and $\beta$ is the angle between the plates[3].

### 9.3.1 Parasitics

Surface-micromachined capacitors will have a number of parasitic components (figure 9.5). The series resistance is due to electrode and lead wire resistance.

The parasitic capacitance, $C_{ox}$, is the capacitance between the bottom electrode and the substrate. The subscript 'ox' comes from microelectronics, where this capacitance would be due to the oxide layer. In surface-micromachining, this capacitance will be due to the isolation layer[4].

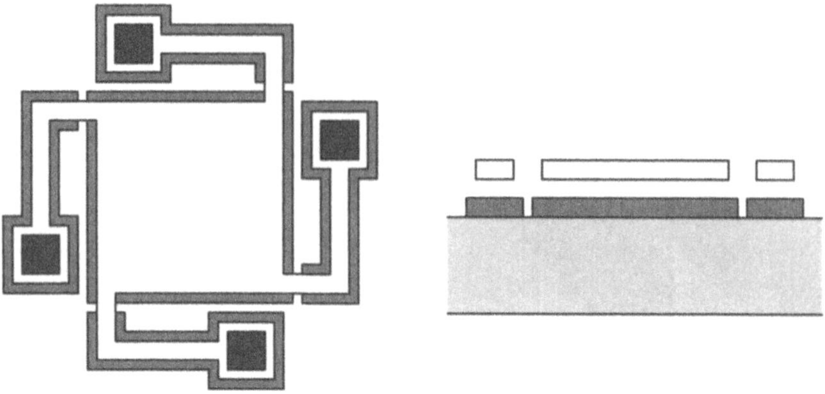

(a) Top-view (b) Side-view

*Figure 9.7.* Illustration of a voltage-tuneable capacitor [9]. In the layout, medium grey represents the ground plane, dark grey represents anchors, and white represents the structural layer.

### 9.3.2 Voltage-Tuneable Capacitors

Voltage-tuneable capacitors can be achieved using a reverse-biased diode[5]. Changing the applied DC voltage changes the size of the depletion region, and so changes the devices small-signal capacitance.

Mechanically, tuneable capacitors can be created by plates whose relative position can be controlled. An example of this approach is displayed in figure 9.7. The applied DC voltage acts to pull the suspended top plate, which is on a flexible suspension, towards the ground plane (section 11.1). This changes the distance between the two plates, and so changes the capacitance.

The capacitor illustrated in figure 9.7 was used to create a variable capacitor[6] [9]. It had a capacitance of 2.2 pF, which could be tuned by 16% for an applied DC voltage of 5.5 volts. At 2 GHz, the capacitor had a $Q$ of 62.

Each plate was 200 by 200 $\mu$m, and the plate separation was 1.5 $\mu$m.

Another design for voltage-tuneable capacitors was investigated by Chinthakindi *et al* [10]. They used large plates which were curved due to non-uniform residual stresses (section 7.3.2). The application of a DC voltage pulled the curved electrodes closer together, increasing their mutual capacitance.

## 9.4 Inductors

Inductors are difficult to manufacture using surface-micromachining. Ideally, one would build a solenoid, but solenoids are difficult to manufacture in a planar process.

Small inductors can be built by laying the conducting wire into a spiral pattern [2,6,11], if a separate conducting path to the centre can be built. However,

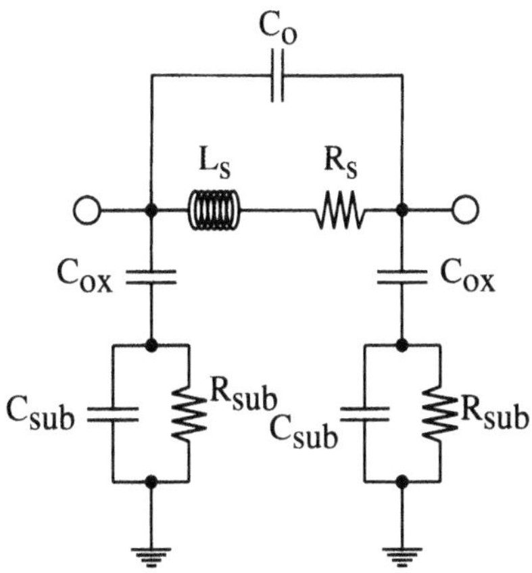


(a) Inductor with substrate parasitics

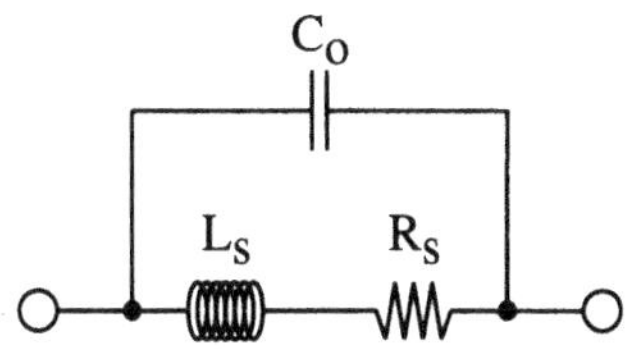


(b) Inductor without substrate parasitics

*Figure 9.8.* Electrical models for non-ideal inductors [6].

parasitic capacitance between the spiral inductor and the substrate lead to self-resonance when the substrate is conducting. This limits the quality factors of spiral inductors to approximately 20 [2]. This in turn limits the maximum operating frequency of the inductor. As the inductance is increased, the substrate capacitance also increases so that the self-resonance frequency decreases [11].

Additionally, resistive losses in the substrate can severely limit the quality factor ($Q$).

Minimizing the substrate capacitance, or using insulating substrates, can alleviate both of the problems just discussed. For example, simulations of a 100 nH inductor[7] showed the self-resonance to be 800 MHz. When the substrate was removed, simulations led to a self-resonance at 3 GHz and a increase in the quality factor [11]. Ziaie *et al* [6] created components useable up to approximately 20 GHz.

Another method of increasing inductor performance is to incorporate magnetic cores [1]. In a planar process, this involves depositing a magnetic material, such as NiFe, under or over the spiral inductor.

For a square spiral inductor, the low-frequency inductance can be determined as follows [6]:

$$L \approx \frac{45\mu_0 n^2 a^2}{22r - 14a} \tag{9.7}$$

In equation 9.7, $L$ is the inductance, $\mu_0$ is the permeability, $n$ is the number of turns, $a$ is the square spiral's mean radius, and $r$ is the outer radius (figure 9.9). Equation 9.7 should be accurate to 5%.

(a) Top-view (b) Side-view (c) Enlarged Side-view

*Figure 9.9.* Structure of a square spiral planar inductor.

The DC parasitic resistance, $R_s$, of the square spiral inductors is easily calculated from the geometry and the material's resistivity (see equation 9.3).

## Notes

1 Off-chip components present additional disadvantages in the extra manufacturing effort required in component assembly (section 1.1).

2 In particular, high-speed electronics require carefully designed and characterized bonding pads.

3 For wedge capacitors that are flat, $\beta = \pi$.

4 The isolation layer is not necessarily oxide. For example, it is silicon nitride in PolyMUMPs$^{TM}$.

5 Diodes designed to be used in this manner are often called varactors.

6 Four devices were connected in parallel.

7 The geometry of the inductor was a square spiral with 20 turns. The lines were 4 $\mu$m wide, and the line spacing was also 4 $\mu$m.

## References

[1] C.T.-C. Nguyen, L.P.B. Katehi, and G.M. Rebeiz. Micromachined devices for wireless communications. *Proceedings of the IEEE*, 86(8):1756–1768, 1998.

[2] C.T.-C. Nguyen and R.T. Howe. An integrated cmos micromechanical resonator high-q oscillator. *IEEE Journal of Solid-State Circuits*, 34(4):440–455, 1999.

[3] D. Koester, A. Cowen, R. Mahadevan, M. Stonefield, and B. Hardy. *PolyMUMPs Design Handbook*. MEMSCAP, 2003.

[4] V. Koeninger, H.H. Uchida, and E. Fromm. Degradation of gold-aluminium ball bonds by aging and contamination. *IEEE Transactions on Components, Packaging, and Manufacturing Technology, Part A*, 18(4):835–841, 1995.

[5] T.C. Wei and A.R. Daud. The effects of aged cu-al intermetallics to electrical resistance in microelectronics packaging. *Microelectronics International*, 19(2):38–43, 2002.

[6] B. Ziaie and K. Najafi. A generic micromachined silicon platform for high-performance rf passive components. *Journal of Micromechanics and Microengineering*, 10(3):365–371, 2000.

[7] J.M. Rabaey, A. Chandrakasan, and Borivoje Nikolic«. *Digital Integrated Circuits: Second Edition*. Prentice-Hall, Upper Saddle River, New Jersey, 2003.

[8] S. Zappe, M. Baltzer, T. Kraus, and E. Obermeier. Electrostatically driven linear microactuators: Fe analysis and fabrication. *Journal of Micromechanics and Microengineering*, 7(3):204–209, 1997.

[9] D.J. Young and B.E. Boser. A micromachined variable capacitor for monolithic low-noise vco's. In *Technical Digest 1996 Solid-State Sensor and Actuator Workshop*, pages 86–89, 1996. Hilton Head Island, SC, June 3-6.

[10] A.K. Chinthakindi, D. Bhusari, B.P. Dusch, J. Musolf, B.A. Willemsen, E. Prophet, M. Roberson, and P.A. Kohl. Electrostatic actuators with intrinsic stress gradient - i. materials and structures. *Journal of the Electrochemical Society*, 149(8):H139–H145, 2002.

[11] J.Y.-C. Chang, A.A. Abidi, and M. Gaitan. Large suspended inductors on silicon and their use in a 2-$\mu$m cmos rf amplifier. *IEEE Electron Device Letters*, 14(5):246–248, 1993.

Chapter 10

# MECHANICAL COMPONENTS

In macroscale systems, a large number of techniques are used to reduce friction, such as ball bearings, magnetic levitation, and hydrostatic bearings [1]. Unfortunately, these techniques either do not work well in the microscale (section 4.1), or are difficult to manufacture. In particular, difficulties with large friction forces (section 4.4) and adhesion (section 4.3.5) fundamentally changes the design problem. This leads to alternate techniques for making mechanical systems at the microscale.

## 10.1 Compliant Devices

Compliant devices are mechanical devices that use their inherent elasticity to perform their function. In contrast, traditional mechanical devices are designed so that the components are stiff and motion is allowed by joints[1]. In compliant devices, mobility is instead allowed by the flexibility of components [2].

Compliant devices are beneficial to manufacturing since they reduce the amount of assembly required. Compliant devices are monolithic, a design philosophy that fits very well with surface-micromachining. Further, compliant devices scale very well. A review of the structures used by microorganisms would reveal heavy reliance on compliant structures [3].

Thus surface-micromachining makes heavy use of flexible beams and diaphragms [3, 4]. Successful designs often replace traditional joints with compliant systems that approximate the joints' functions. For example, suspensions replace prismatic joints[2] and flexure hinges replace revolute joints[3].

The mechanical structures discussed in this chapter will all be compliant structures. These obviously form a small subset of all possible compliant mechanisms. While the design of new compliant structures is certainly possible, design methods typically rely on numerical optimization methods [3].

*Figure 10.1.* SEM of a device that uses both sliding joint. and compliant joints. Fabricated using PolyMUMPs$^{TM}$.

The design of novel compliant structures will not be discussed. Instead, readers are referred to the literature [5–7].

The above is not to say that traditional prismatic and revolute joints are never used. They can be built in surface-micromachining technologies, and while compliant systems are preferred, traditional sliding contact joints are not rare. The main advantage of traditional joints is the elimination of restoring forces. Thus, devices using sliding contact joints can have longer ranges of travel.

A good example of the use of compliant devices can be seen in figure 10.1. First, the actuators are compliant devices (section 12.2). The actuators are connected together into teams using compliant beams. Horizontal motion of the teeth in the upper right corner is provided by compliance in the beam. Despite all the mechanical links, there is only one sliding contact joint in this device, the prismatic joint providing vertical motion of the output teeth.

## 10.2 Springs

Springs are simply mechanical device's that obey Hooke's law.

$$F = kx \tag{10.1}$$

Above, $F$ is the force necessary to deform a spring by distance $x$. The constant, $k$ is known as the stiffness. However, equation 10.1 is only for one dimensional systems.

A surface-micromachined spring, which is a two-dimensional spring, has additional forces and displacements to consider. To completely characterise a two-dimensional spring, one must determine the relations between forces and displacements along both axes, as well as torque and rotation. Thus, in two dimensions, the stiffness can be expressed as a $3 \times 3$ matrix.

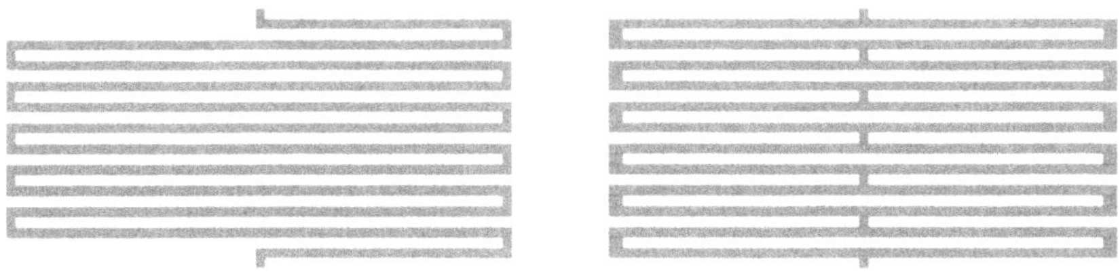

(a) Serpentine Spring (b) Box Spring

*Figure 10.2.* Layouts for simple springs. Connections to these springs are made at both the top and bottom.

$$\vec{F} = \mathbf{K}\vec{x} \tag{10.2a}$$

$$\vec{F} = [F_x \quad F_y \quad M]^T \tag{10.2b}$$

$$\vec{x} = [x \quad y \quad \theta]^T \tag{10.2c}$$

Despite the multiple components in equation 10.2a, the assumption is that a mechanism described as a spring has the stiffness specified for a single motion, while the other degrees of freedom are unspecified.

### 10.2.1 Serpentine Spring

In surface-micromachining, the easiest method of creating a spring is to lay out a serpentine path (figure 10.2a).

The axial compliance of a serpentine spring can be calculated using the following formula:

$$C_y = \frac{L^3}{EI}\left(\frac{N}{12} + \frac{1}{48}\right) \tag{10.3}$$

Where, $C_y$ is the compliance of the spring, $L$ is the length of the bars, $E$ is Young's Modulus for the spring's material, $I$ is the bending moment of inertia for the bars, and $N$ is the number of bars, excluding the two half-bars at the top and bottom of the spring. The two half-bars, which are necessary so that connections are made along the spring's axis, are included in the above equation by the fraction $1/48$.

The bar length, $L$, is not the distance between the outside edges of the spring. It is the distance between the axes of the short beams connecting the bars together. If the serpentine path is laid out with a width $w$, then the distance between the outside edges is $L + w$, since $w/2$ is picked up at each edge.

The compliance is the inverse of the stiffness. To use in Hooke's law (equation 10.1), the stiffness is simply $k_y = C_y^{-1}$.

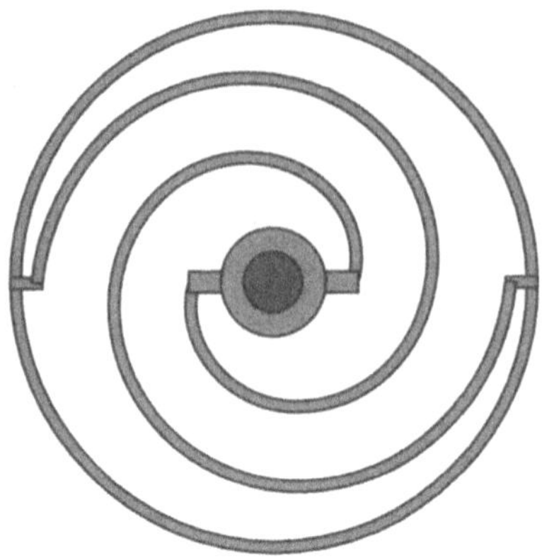

*Figure 10.3.* Layout for a spiral or torsional spring. The dark region in the centre is where the spring is anchored to the substrate.

Equation 10.3 makes several approximations, the most important probably being that the bending in the sections connecting the bars together is negligible. As long as these connections are short compared to the bars, the approximation will hold. Another method to ensure this approximation is valid would be to increase the width of the connecting bars. If bending of the connecting bar is not negligible, then equation 10.3 will underestimate the compliance.

### 10.2.2 Box Spring

Instead of a serpentine path, the spring can be constructed from a number of connected boxes (figure 10.2b). This type of spring is considerably stiffer than a serpentine spring for the same area. However, it does not have the same tendency to rotate or move sideways under axial loads because it is symmetric. The box spring is therefore more suitable in situations where the components must remain axially aligned.

Unfortunately, box springs are stiffer.

$$C_y = \frac{NL^3}{96EI} \tag{10.4}$$

The variables in equation 10.4 are the same as in equation 10.3, except for $N$. For the box springs, $N$ refers to the total number of boxes.

Again, equation 10.4 assumes that no bending occurs in the segments connecting the horizontal beams together.

### 10.2.3 Spiral Spring

Spiral springs are used to create rotary or torsional springs. The geometry of spiral springs usually follows an Archimedean spiral, which is where the spiral's radius is linearly proportional to the angle, as follows:

$$r(\theta) = c\theta \tag{10.5}$$

In equation 10.5, the radius of the spiral, $r$, is determined from the angle, $\theta$, and the proportionality constant, $c$

Whereas serpentine and box springs are designed to be loaded axially, spiral springs are designed to be loaded with a moment. Ideally, their displacement is a rotation that is proportional to the applied moment ($M = k\theta$). Thus, spiral springs have a torsional stiffness, which can be calculated as follows:

$$k = \frac{EI}{L} \tag{10.6a}$$

$$= \frac{hw^3E}{12L} \tag{10.6b}$$

In equation 10.6a, $E$ is Young's Modulus for the spring's material, $I$ is the bending moment of inertia for the bars, and $L$ is the total length of the sprial. If the spiral's beam has a rectangular cross-section, the the moment of inertia can be expanded, which results in equation 10.6b. The new variables are $w$, the width of the beam, and $h$, the height of the beam.

## 10.3 Suspensions

Suspensions are similar to springs. Suspensions are designed to be compliant and so readily stretch when a force is applied. However, suspensions are ideally compliant along only one axis [8]. In response to perpendicular forces or applied moments, they should be very stiff. An ideal suspension would provide a single degree of freedom for movement. Suspensions are designed to replace prismatic joints.

In reality, even in the allowed degree of freedom, suspension will present some degree of stiffness. In conventional mechanical components, they are better represented by a joint and a spring[4]. To increase accuracy, modelling must include multiple ideal joints and springs to represent all possible motions and restoring forces[5].

Suspensions are similar to flexure joints (section 10.4) in that they are designed to mimic an ideal mechanical joint. However, suspensions differ in that they use multiple beams to constrain the motion of a central component, usually called the shuttle.

### 10.3.1 Clamped-clamped Suspension

The simplest type of suspension is the clamped-clamped suspension (figure 10.4). The shuttle's motion is constrained by two beams, which are anchored at the outside. The restoring forces on the shuttle are easily calculated from simple beam theory (chapter 5). The axial stiffnesses are [8,9]:

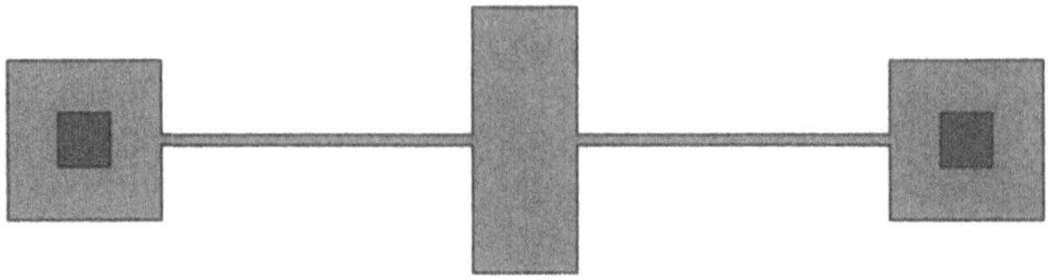

*Figure 10.4.* Illustration of a clamped-clamped suspension. The ends are anchored to the substrate, as indicated by the dark regions.

$$\begin{aligned} k_x &= \frac{2EA}{L} \\ &= \frac{2Ewh}{L} \end{aligned} \tag{10.7}$$

$$\begin{aligned} k_y &= \frac{24EI_z}{L^3} \\ &= \frac{2Ew^3h}{L^3} \end{aligned} \tag{10.8}$$

In equation 10.7, $E$ is Young's modulus for the suspension's material, $A$ is the cross-sectional area of the beams, $L$ is beams' length, $w$ is the beams' width, and $h$ is the beams' height. In equation 10.8, the moment of inertia, $I_z$, has also been used.

The stiffness ratio of clamped-clamped suspensions can be very high [8]. For example, a suspension with beams that are 500 $\mu$m long and 2 $\mu$m wide will have a stiffness ratio[6] of $1.25 \times 10^5$.

Unfortunately, if the shuttle experiences large displacements, then Hooke's law may no longer apply, limiting the accuracy of equations 10.7 and 10.8. The beams will experience stress stiffening, which will increase the beams' stiffnesses as the deflection increases. Stress stiffening is a non-linear effect, leading to errors in Hooke's law. A more complete equation for the deflection-force relationship can be found in several references [8, 9]. Hooke's law is accurate for deflections only up to one quarter of the width of the beams. This is a very severe limitation, and the primary reason why clamped-clamped suspensions are not much used in practice.

Additionally, the behaviour of clamped-clamped suspensions is very sensitive to residual stresses (section 7.3). Through stress stiffening, residual stresses can cause large changes in suspensions' stiffnesses. Further, if sufficient compressive stress is present, the suspension may buckle (section 5.4.1).

Folded-beam suspensions, also discussed this chapter, allow residual stresses to relax, and so prevent the difficulties associated with residual stresses that clamped-clamped suspensions present.

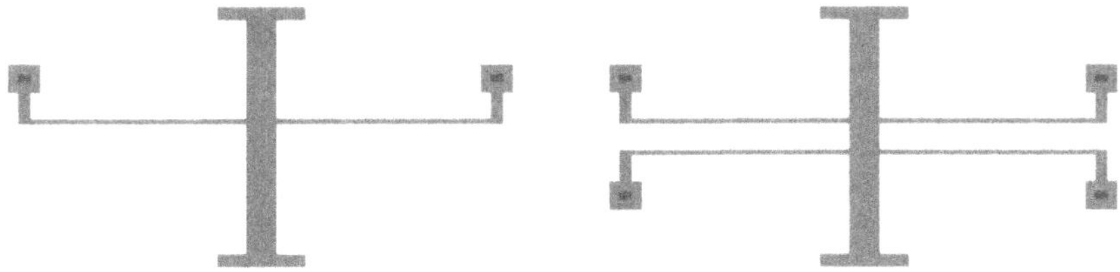

(a) Crab-leg suspension with one beam on each side.

(b) Crab-leg suspension with two beams on each side.

*Figure 10.5.* Layouts for two types of crab-leg suspensions.

### 10.3.2 Torsional Spring

A clamped-clamped suspension can also be used as a torsional spring. The axis of rotion is along the beams. Please refer to section 5.2.

### 10.3.3 Crab-leg Suspension

Another suspension, which reduces the axial forces compared to the clamped-clamped suspension, is the crab-leg suspension (figure 10.5a). The outside end of each beam has another section parallel to the shuttle. The inside beams, which are perpendicular to the shuttle, are known as the thigh segments, and the parallel or outside beams are known as the shin segments [8]. Note that as the shin length approaches zero, then the stiffnesses of crab-leg suspensions approach that of clamped-clamped suspensions.

The stiffness of the crab-leg suspension illustrated in figure 10.5a can be determined as follows [8]:

$$k_x = \frac{12EI_2}{L_2^3}\left(\frac{L_1I_2 + 2L_2I_1}{2L_1I_2 + L_2I_1}\right) \tag{10.9}$$

$$k_y = \frac{24EI_1}{L_1^3}\left(\frac{L_1I_2 + L_2I_1}{L_1I_2 + 4L_2I_1}\right) \tag{10.10}$$

In equations 10.9 and 10.10, $L_1$ and $I_1$ refer to the length and moment of inertia of the thigh, while $L_2$ and $I_2$ refer to the length and moment of inertia of the shin.

The crab-leg suspension illustrated in figure 10.5b has two sets of beams on each side of the shuttle. Only one set of beams on each side is necessary, but like the folded-beam suspension, the extra set provides coupled forces, and helps prevent rotation of the shuttle. This structure is stiffer, axially, transversely, and rotationally. This structure is used when one needs to not only provide a large stiffness ratio between $k_x$ and $k_y$, but also a large stiffness ratio between $k_\theta$ and $k_y$.

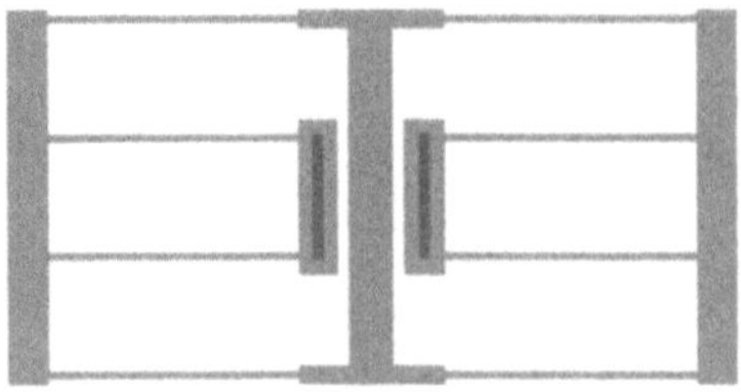

(a) Folded-beam Suspension

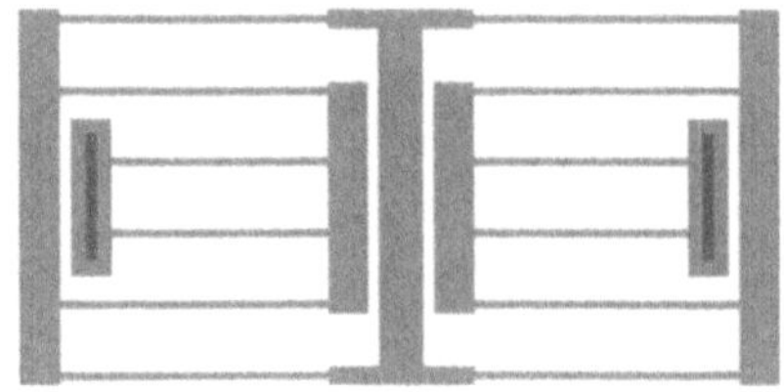

(b) Folded-beam Suspension (with additional fold)

*Figure 10.6.* Layout of two folded-beam suspensions. In (a) the beams are folded once while in (b) the beams are folded twice.

For crab-leg suspensions as illustrated in figure 10.5b, the stiffnesses can be calculated using equations 10.9 and 10.10. However, an additional factor of two is required.

The crab-leg suspension increases the linear travel range, but reduces the stiffness ratio [8]. For example, the addition of shins with lengths $L_2$ equal to 50 $\mu$m to the example clamped-clamped suspension discussed earlier, reduces the stiffness ratio by several orders of magnitude [8].

Legtenberg *et al* [8] produced crab-leg suspensions with measured stiffnesses 17 $\mathrm{Nm}^{-1}$ and 0.029 $\mathrm{Nm}^{-1}$ for $k_x$ and $k_y$, respectively. They predicted values showed significant errors, which were attributed to non-rectangular beams. Unfortunately, the stiffnesses are very sensitive to beam geometry, making good control of the stiffnesses difficult (section 7.2).

### 10.3.4 Folded-beam Suspension

In a folded-beam suspension, the shuttle is connected to four straight beams that are oriented perpendicular to the direction of motion. For loads aligned with the shuttle's axis, the beams are loaded transversely, where they are compliant. However, for loads perpendicular to the shuttle's axis, the beams are loaded axially, and so are stiff. Thus, depending on the direction of the applied load, a folded-beam suspension can be either stiff or compliant.

Four beams are used so that there are two beams on each side of the shuttle. The paired beams on each side can, through axial forces, provide a force couple. These force couples create a moment, which can resist any applied moments. This provides stiffness against rotation.

In a folded-beam suspension, the beams are folded in half to save space (figure 10.6a). The flexing occurs along two beams of length $L/2$ instead of one beam of length $L$. The single beam would be more flexible, but would also occupy more space.

The usual configuration is to fold the beams once (figure 10.6a). However, the beams can be folded multiple times (figure 10.6b).

The compliance of the folded-beam suspension can be determined using the following equation [8, 10]:

$$k_x = \frac{2EA}{L} \tag{10.11}$$

$$k_y = \frac{24EI_z}{L^3} \tag{10.12}$$

In equation 10.11, $k_x$ is the stiffness along the the x-axis, $E$ is Young's modulus for the suspension's material, $A$ is the cross-sectional area of the beams, and $L$ is the beam's length. In equation 10.12, $k_y$ is the stiffness along the the y-axis[7], $L$ is the length of the beams, $E$ is Young's modulus, and $I_z$ is the moment of inertia for bending about the z-axis.

The stiffness of the device along the z-axis, $k_z$ is similar to equation 10.12, except that $I_z$ is replaced by $I_x$.

The above equations assume that the other sections of the suspension are rigid. When creating the layout, it is therefore important that the other sections are considerably wider than the bending beams.

Folded flexure suspensions allow significant relaxation of axial forces, and so Hooke's law applies for much larger deflections that previously discussed suspensions [8]. However, the stiffness ratio is the same as for clamped-clamped suspensions. This does not mean that non-linear effects are not present. More complete expressions for the deflection-force relationship can be found in Legtenberg *et al* [8]. These show that Hooke's law applies for deflections up to approximately 10% of the beam length.

Despite the previous statement, important non-linear affects occur before displacement reaches $0.1L$. As the shuttle is displaced, the transverse spring constant, $k_x$, decreases. This also leads to smaller stiffness ratio. Details of this calculation can be found in Legtenberg *et al* [8].

The resonance frequency of the suspension can be calculated as follows [8, 10]:

$$\omega_0 = \sqrt{\frac{k_y}{M_{eq}}} \tag{10.13}$$

In equation 10.13, the resonance frequency, $\omega_0$, is calculated from the axial stiffness, $k_y$, and the equivalent mass, $M_{eq}$. The equivalent mass is related to, but less than, the total mass of the suspension. The density is further weighted by the relative motion. For folded-beam suspension, it can be approximated by [10]:

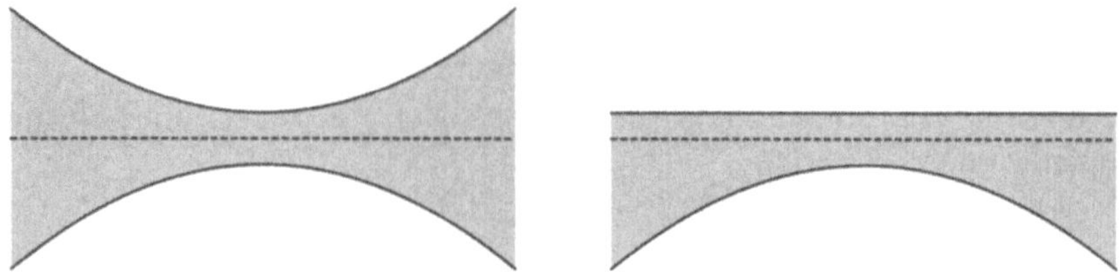

(a) Symmetric Flexure Hinge (b) Non-symmetric Flexure Hinge

*Figure 10.7.* The difference between symmetric and non-symmetric flexure hinges.

$$M_{eq} = M_{load} + M_{shuttle} + 2.73M_{beam} + 0.5M_{stiff} \qquad (10.14)$$

In equation 10.14, $M_{load}$ is the mass of rigid bodies connected to the shuttle[8], $M_{shuttle}$ is the mass of the shuttle, $M_{beam}$ is the mass of the beams, and $M_{stiff}$ is the mass of the connecting beams at the ends of the beams.

Folded-beam suspensions are commonly used with interdigitated comb-drives (section 11.2).

## 10.4 Flexure Joints

Flexure joints are compliant devices designed to replace traditional sliding-contact joints [11]. Depending on the design, a flexure joint may be a replacement for a revolute joint, a prismatic joint, or a combination. They are thus similar to suspensions. However, flexure joints connect to parts using a single connecting piece.

### 10.4.1 Revolute Joints

Flexure joints which are specifically designed to replace revolute joints are often commonly referred to as flexure hinges. The design of flexure hinges, in particular, has received much attention. They are valuable in traditional[9] machining where they can eliminate assembly [2] and lead to increased precision [2].

Flexure joints are typically narrow sections connecting two larger pieces. Stiffness to bending depends on the bending moment of inertia, which scales as $s^3$ for two dimensional structures[10]. Thus, narrow sections will have significantly higher compliance to bending. Properly designed, flexure hinges can closely approximate revolute joints.

Flexure joints come in several basic forms [12]. However, the constraints imposed by surface-micromachining reduce the available joints in surface-micromachined mechanisms to one basic class: two-dimensional or planar flexure hinges. This class of flexure hinges can be further broken down into symmetric and non-symmetric hinges (figure 10.7).

| Type | Equation |
|---|---|
| Rectangular | $t(x) = t_0$ |
| Circular | $t(x) = t_0 + 2\left[r - \sqrt{x(2r-x)}\right]$ |
| Corner-Filleted | $t(x) = \begin{cases} t_0 + 2\left[r - \sqrt{x(2r-x)}\right] & , x \in [0, r] \\ t_0 & , x \in [r, l-r] \\ t_0 + 2\left[r - \sqrt{(l-x)(2r-l+x)}\right] & , x \in [l-r, l] \end{cases}$ |
| Parabolic | $t(x) = t_0 + 2c\left(1 - 2\frac{x}{l}\right)^2$ |
| Hyperbolc | $t(x) = \sqrt{t_0^2 + 4c(c+t_0)\left(1 - 2\frac{x}{l}\right)^2}$ |
| Inverse Hyperbolic | $t(x) = \frac{2a}{b^2-(x-l/2)^2}, a = \frac{tl^2}{8}\left(1+\frac{t}{2c}\right), b = \frac{l}{2}\sqrt{1+\frac{t}{2c}}$ |
| Secant | $t(x) = t/\cos\left(\frac{2x-l}{l}\arccos\frac{t}{2c+t}\right)$ |

*Table 10.1.* Several types of planar flexure hinges [12].

Even this limited class of flexure hinges leaves a fair bit or room for design. Design of flexure hinges can be viewed as selecting the profile for the perimeter of the flexure hinge. For example, the simplest flexure hinge has a rectangular cross-section. The perimeter of this flexure hinge can be described as follows:

$$t(x) = t_0 \tag{10.15}$$

The shape of a flexure hinge is given by a function $t(x)$, which, for a rectangular cross-section, is given by equation 10.15. The function $t$ is the distance to the edge from the axis, where $x$ is the distance along the axis. For a rectangular hinge, this is a fixed distance, $t_0$.

For a symmetric hinge, the bottom edge is can be determined by using $-t(x)$. For a non-symmetric hinge, the bottom is given by $-t_0$. Note that for a rectangular hinge, there is no difference between the symmetric and non-symmetric types.

While other profiles are possible, the flexure hinges listed in table 10.1 have the advantage of closed-form solutions for their compliance matrices, which can be obtained from [12].

## 10.4.2 Beam Joints

Beam joints connect two rigid[11] bodies together. This is similar to the revolute joints just discussed. However, instead of varying the profile of the beam, beam joints use a beam with a fixed cross-section, but different paths. The design of beam joints thus shifts focus from the beam profile to the beam path.

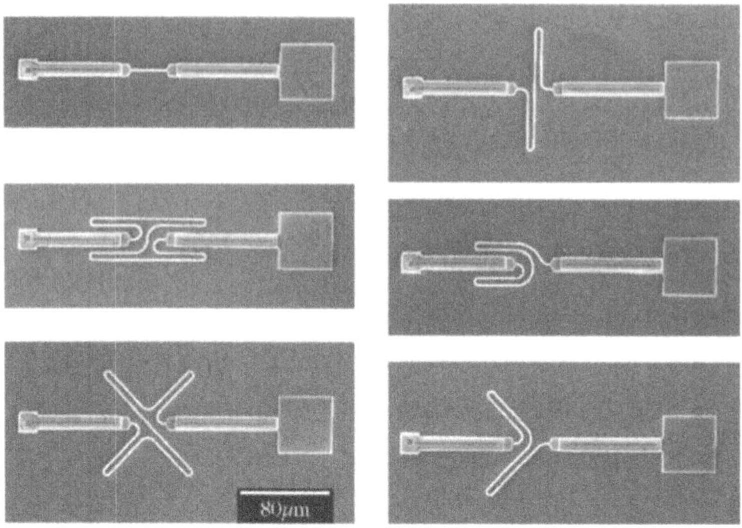


*Figure 10.8.* SEMs of several flexure joints. These are the 'I'-joint, 'H'-joint, 'X'-joint, 'S'-joint, 'U'-joint, and 'V'-joint [11]. Copyright © 2001 IOP.

The simplest beam hinge is a rectangular cross-section joint, connected in a straight line between the two rigid bodies. This joint, also known as an 'I'-joint, is thus the same the rectangular joint discussed in the last section. The simplest flexure joint thus belongs to both categories.

To increase the compliance of this joint, the beam must be lengthened. However, this would also increase the necessary spacing between the two rigid bodies. Instead, the beam is folded into the available space. Depending on how the beams are oriented to the two rigid bodies, different types of joints can be approximated. Beams are stiffer axially than they are transversely. The 'H'-joint (figure 10.8) therefore approximates an axial prismatic joint while the 'S'-joint approximate a transverse prismatic joint.

The torsional compliance is proportional to the total length of beam connecting the two rigid bodies. By varying the orientation and total length of beams, the different compliances of the $3 \times 3$ compliance matrix can be controlled.

## 10.5 Levers

A lever is a device that scales an input force or displacement. Levers normally consist of four parts: the lever arm, the fulcrum or pivot, the input and output.

In an ideal lever, the output displacement is a multiple of the input displacement.

$$d_{out} = d_{in} \times GA \tag{10.16}$$

Above, the output displacement, $d_{out}$ is related to the input displacement, $d_{in}$ and the geometric advantage, $GA$. Because work must conserved, the output force must also be scaled. This following result also follows from balancing the moments created by the forces.

$$F_{out} = F_{in} \times MA \tag{10.17a}$$

$$F_{out} = F_{in} \times \frac{1}{GA} \tag{10.17b}$$

The output force is related to the input force through the mechanical advantage, $MA$. In ideal levers, the force advantage is the inverse of the geometric advantage. The mechanical advantage is essentially the force amplification, while the geometric advantage is the displacement amplification.

The above only applies to ideal levers. With pivot or revolute joints that have negligible friction or stiffness, levers that closely approximate the above equations can be built. However, in micromachines, surface-contact joints[12] are often replaced by flexures. In these cases, some of the work done at the input is absorbed in the elasticity of the the lever's joints. In levers using flexure joints, the following relation must hold.

$$GA \times MA \leq 1 \tag{10.18}$$

The product of the geometry advantage and the force advantage is a good figure of merit for flexure based levers.

Levers can be classified by the location of the fulcrum compared to the location of the input and output forces [4]. For a class 1 lever, the fulcrum is between the input and output forces, for a class 2 lever, the fulcrum is located past the output[13], and for a class 3 lever, the fulcrum is past the input[14].

Class 2 levers are typically used to amplify force, while class 3 levers ar typically used to amplify displacement [4]. Class 1 levers can either amplify force or displacement, depending on the distances between the output and input.

For ideal levers, the geometric and mechanical advantage can be calculated as follows:

$$GA = \frac{L_{out}}{L_{in}} \tag{10.19a}$$

$$FA = \frac{L_{in}}{L_{out}} \tag{10.19b}$$

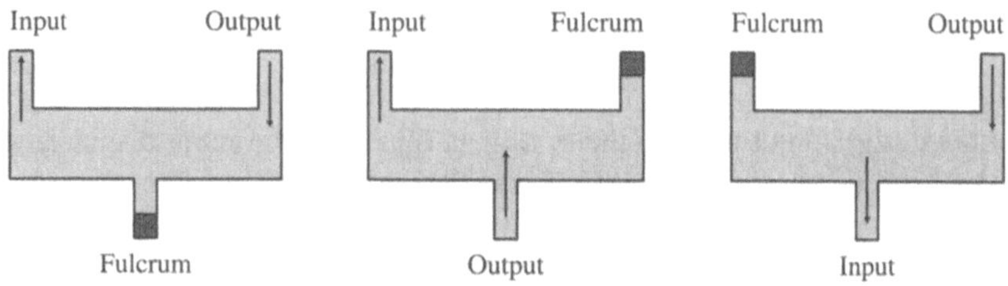


(a) Class 1 (b) Class 2 (c) Class 3

*Figure 10.9.* Illustration of the three classes of levers. Designs are for microlevers.

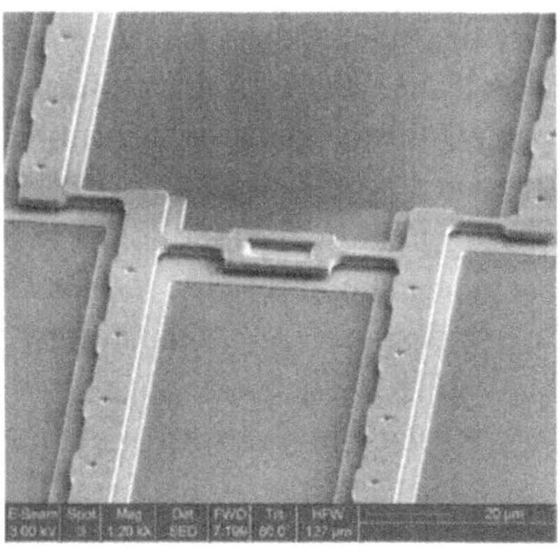

*Figure 10.10.* SEM of a pair of opposing microlevers. Fabricated using PolyMUMPs™.

In equation 10.19a, $L_{out}$ and $L_{in}$ are the distances between the fulcrum and the output and input connections, respectively.

The following section discusses microlevers, which are compliant implementations of traditional levers. They have readily identifiable lever arms. However, there are other systems which are capable of providing force or geometry advantage that do no have lever arms. The goal, with all these devices, is to design a mechanism with the desired force or geometric advantage while maximizing $GA \times MA$ 10.18.

## 10.5.1 Microlevers

Microlevers have a ridgid bar to act as the lever arm, but differ from traditional levers in that the fulcrum, input, and outputs are all compliant (figure 10.9).

In addition to being split up into class 1, 2, and 3, microlevers are futher classified by whether the output and fulcrum are on the same side of the lever arm [4, 13]. The side that the input is connected to is not important. This leads to six classes of microlevers[15].

The force advantage of a microlever, under certain conditions, can be determined using the following equation [4, 13]:

$$MA = \frac{(1/k_{vvp})(k_{\theta mo} + k_{\theta mp}) \pm L_{in}L_{out}}{(1/k_{vvo} + 1/k_{vvp})(k_{\theta mo} + k_{\theta mp}) + L_{in}^2} \qquad (10.20)$$

In equation 10.20, a number of stiffness terms have been introduced. For example, $k_{vvo}$ is the vertical stiffness for an applied vertical force for the output connection, and $K_{vvp}$ is the vertical stiffness for an applied vertical force of the pivot. The first subscript can also be $\theta$, referring to rotation. The second subscript can also be $m$, referring to an applied moment.

For equation 10.20 to be valid, certain conditions need to be met [13]. These are (1) that all deformations are linear, (2) that the lever arm is rigid, (3) that the orientation of the ends of the fulcrum beam and the output beam will remain fixed after loading, (4) that the vertical output force does not cause any horizontal or rotational displacement of the output system, and (5) that the horizontal forces on the pivot and output beams are negligible. Su and Yang [13] also provide equations that rely only on the first four conditions. This model allows for horizontal forces on the fulcrum and output beams.

Of interest is the ratio between the ideal mechanical advantage and the actual mechanical advantage, as determined by equation 10.20. Assuming that $L_{in} \ll L_{out}$, then this ratio can be written as [4]:

$$\frac{MA}{MA_0} \approx \frac{L_{in}^2}{(1/k_{vvo} + 1/k_{vvp})(k_{\theta mo} + k_{\theta mp}) + L_{in}^2} \qquad (10.21)$$

This leads to the definition of the microlever amplification coefficient [4]:

$$MA^* = \left(\frac{1}{k_{vvo}} + \frac{1}{k_{vvp}}\right)(k_{\theta mo} + k_{\theta mp}) \qquad (10.22)$$

When the amplification coefficient increases, the actual mechanical advantage decreases. Thus, good microlever designs should seek to minimize $MA^*$. The goal is to have the pivot and output systems axially stiff, but rotationally compliant.

Furthermore, under the assumption that the output is nearly ideal, then $MA^*$ reduces to $w_p^2/12$, where $w_p$ is the width of the pivot beam[16]. This clearly indicates the need to minimize the width of the fulcrum's beam. Similar result show that the width of the output beam, $w_o$, is also critical and should be minimized.

The input and output axial compliances of the lever should also be selected to match the mechanical systems connected to the input and output. Typically,

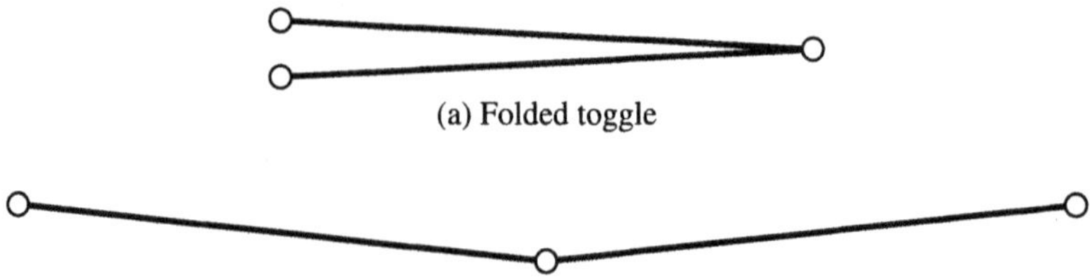


*Figure 10.11.* Illustration of the two types of toggle devices. In both diagrams, lines represent links while circles represent revolute joints.

the best performance is achieved by matching the compliances. A more detailed explanation of the design of multi-stage microlevers is detailed by Su and Yang [4].

### 10.5.2 Toggle Mechanisms

Toggle mechanisms are a class of compliant devices capable of providing geometry amplification. It is interesting to note that these devices are very similar in topology to Guckel actuators (section 12.2) and bent-beam actuators (section 12.3). Further, the open toggle mechanism, to be introduced shortly, is also central to an important class of strain sensors (section 7.3).

When two rigid links of a device are collinear[17], then the device is in toggle [14]. This takes two forms (figure 10.11). The collinear links are either side by side, called a folded toggle, or end to end, called an open toggle. In both cases, the links are connected with a revolute joint. The open toggle, when used as a compliant device, is often referred to as a bent-beam.

To achieve mechanical advantage, one of the end links is fixed[18], and a displacement is applied to the other end. The output is taken from the joint connecting the two links.

If the input displacement is aligned with the input link, then the system can be modelled as a change in the link length. This leads to a picture of the joint motions expressed in figure 10.12. With the input displacement modelled as a change in link length, then only joint C moves, and its final position is C'. If the length of the link AC, initially $l_0$, is lengthed by displacement $\mathrm{d}l$, the joint C will be displaced a distance $\mathrm{d}x$ to new position C'. The geometric advantage is, by definition, then [14]:

$$GA = \frac{\mathrm{d}x}{\mathrm{d}l} \tag{10.23}$$

For the folded toggle (figure 10.12a), the variable $R$ is the length of the link BC, and the variable $e$ is the perpendicular distance from the joint B to the line through the axis of link AC. From trigonometry, the length of link AC can be

*Figure 10.12.* Relationships used to determine geometry advantage of toggles.

calculated using $l^2 = R^2 + e^2 - 2Re\cos\beta$, with $\beta$ the angle $\angle$ABC. Further, the joint C moves in a circle about joint B, and so $\mathrm{d}x = R\mathrm{d}\beta$. Differentiating to find $\mathrm{d}l$, and simplifying based on the knowledge that $e \ll R$, then the geometry advantage can be found to be [14]:

$$GA = \frac{R}{e} \tag{10.24}$$

In figure 10.12b, the variables $R$ and $e$ are also defined. In this case, $R$ is half of distance between joints A and B, while $e$ is the distance of joint C to the line AB. The lengths $x$ and $l$ are easily found from the angle $\beta$, which is this case is half the angle $\angle$BAC. We thus have $x = l\sin\beta$ and $R = l\cos\beta$. These can be rearranged to find $\mathrm{d}x/\mathrm{d}l$. When $\beta$ is small, the ratio can be simplified with a result equal to equation 10.24 [14].

Typical dimensions for folded toggles are lengths of 200 $\mu$m and distance between the link's axes of 4 $\mu$m [14]. This results in a mechanism with geometric advantage of 50. Alternatively, an open toggle mechanism is typically 400 $\mu$m long, and has a 6° angle [14]. This results in a mechanism with a geometric advantage of 10. In general, open toggle mechanisms have angles in the range of 3° to 6°.

Equation 10.24 assumes ideal revolute joints. In reality, these joints are compliant devices and have non-negligible stiffness. These stiffnesses will reduce the motion at the output, and so equation 10.24 is an upper limit on the geometric advantage. The actual geometric advantage for the folded toggle mechanism described in the previous paragraph was only approximately 60% the predicted value.

One advantage of the folded toggle over the open toggle is that it allows residual stresses (section 7.3) to relax. The output joint moves a distance $\epsilon L$, where $\epsilon$ is the strain and $L$ is the length of the toggle mechanism. An open toggle mechanisms will amplify any residual strain leading to an initial displacement of the output joint of $\epsilon L \times MA$. Open toggles, therefore, make good strain gauges [15].

Gianchandani [15] derived equations for the motion of output joint of an open toggle mechanism, with the input being a force instead of a displacement. The applied force is applied along the axis AB. When the force, $F$, is tensile, then the displacement of the output joint can be determined using the following equation [15]:

$$\Delta y = 2\frac{e}{Rk}\tanh\frac{kR}{2} \tag{10.25}$$

with,

$$k = \sqrt{\frac{F}{EI}} \tag{10.26}$$

If the force is compressive, then the displacement can be found using the following equation [15]:

$$\Delta y = 2\frac{e}{Rk}\tan\frac{kR}{2} \tag{10.27}$$

From the previous two equations, it can be seen that the deflection goes to infinity as the $kR$ approaches $\pi$. This indicates the onset of buckling, and the critical compressive force can be determined to be [15]:

$$F_{cr} = \frac{EI\pi^2}{R^2} \tag{10.28}$$

Gianchandani [15] also provides expressions for the displacement of joint A under an applied force.

## Notes

1 These joints contain surfaces in sliding contact, which is an important reason to avoid them in surface-micromachining.

2 A prismatic joint allows a single linear degree of freedom of motion.

3 A revolute joint allows a single rotary degree of freedom of motion.

4 One figure of merit for suspensions is the ratio of the axial and transverse stiffnesses. Other ratios are also possible.

5 Suspensions approach the ideal as the ratio $k_x/k_y$ is maximized.

6 Note that several parameters cancel when calculating the stiffness ratio.

7 This is the axial compliance.

8 Alternatively, the equivalent mass of any deforming bodies connected to the suspension.

9 i.e. macroscopic.
10 For three dimensional structures, $I \propto s^4$.
11 Or approximately rigid.
12 Such as pivot and revolute joints made from physical separate pieces.
13 The output is between the fulcrum and the input.
14 The input is between the fulcrum and the output.
15 The classes are: 1S, 1D, 2S, 2D, 3S, and 3D.
16 This approximation also depends on the beam having a rectangular cross-section.
17 Clearly, a device cannot be perfectly in toggle, or the two links would occupy the same space.
18 Equivalently, symmetric displacements can be applied to the open toggle.

# References


[1] N.R. Tas, C. Gui, and M. Elwenspoek. Static friction in elastic adhesion contacts in mems. *Journal of Adhesion Science and Technology*, 17(4):547–561, 2003.

[2] L.L. Howell. *Compliant Mechanisms*. John Wiley & Sons, Inc., New York, 2001.

[3] S. Kota, J. Hetrick, Z. Li, and L. Saggere. Tailoring unconventional actuators using compliant transmissions: Design methods and applications. *IEEE/ASME Transactions on Mechatronics*, 4(4):396–408, 1999.

[4] X.-P.S. Su and H.S. Yang. Design of compliant microleverage mechanisms. *Sensors and Actuators A*, 87(3):146–156, 2001.

[5] U.D. Larsen, O. Sigmund, and S. Bouwstra. Design and fabrication of compliant micromechanisms and structures with negative poisson's ratio. *Journal of Microelectromechanical Systems*, 6(2):99–106, 1997.

[6] O. Sigmund. On the design of compliant mechanisms using topology optimization. *Mechanics of Structures and Machines*, 25(4):493–524, 1997.

[7] M.S. Baker and L.L. Howell. On-chip actuation of an in-plane compliant bistable micromechanism. *Journal of Microelectromechanical Systems*, 11(5):566–573, 2002.

[8] R. Legtenberg, A.W. Groeneveld, and M. Elwenspoek. Comb-drive actuators for large displacements. *Journal of Micromechanics and Microengineering*, 6(3):320–329, 1996.

[9] W. Kuehnel. Modelling of the mechanical behaviour of a differential capacitor acceleration sensor. *saa*, 48(2):101–108, 1995.

[10] W.-H. Juan and S.W. Pang. High-aspect ratio si vertical micromirror arrays for optical switching. *jmems*, 7(2):207–213, 1998.

[11] H. Fettig, J. Wylde, T. Hubbard, and M. Kujath. Simulation, dynamic testing and design of micromachined flexible joints. *Journal of Micromechanics and Microengineering*, 11(3):209–216, 2001.

[12] N. Lobontiu. *Compliant Mechanisms: Design of Flexure Hinges*. CRC Press, Boca Raton, 2002.

[13] X.-P.S. Su and H.S. Yang. Analytical modeling and fem simulations of single-stage microleverage mechanism. *International Journal of Mechanical Sciences*, 44(11):2217–2238, 2002.

[14] Y. Lai, J. McDonald, M. Kujath, and T. Hubbard. Force, deflection and power measurements of toggled microthermal actuators. *Journal of Micromechanics and Microengineering*, 14(1):49–56, 2004.

[15] Y.B. Gianchandani. Bent–beam strain sensors. *Journal of Microelectromechanical Systems*, 5(1):52–58, 1996.

Chapter 11

# ELECTROSTATIC ACTUATORS

Electrostatic actuators rely on the force between two conducting electrodes when a voltage is applied between them. Depending on the arrangement of the electrodes, various types of actuators are possible. However, whatever the electrode arrangement, if there exists mutual capacitance, there will also be an attractive force when a voltage is applied between the electrodes. This force will nearly always be attractive[1].

Electrostatic forces are a favourite for surface-micromachined structures. This stems in part from ease of implementation and in part from favourable scaling laws for electrostatic forces (section 4.1) [1]. Electrostatic forces are surface forces, hence their favourable scaling. Further, extremely large electric fields can be created with modest voltage because of the small distances involved [1].

Electrostatic actuators are easy to implement in surface micromachining since structural materials are often conducting. Paired with small manufacturable gaps between components, large electric fields are easily realizable.

One of the main disadvantages of electrostatic actuators is the large voltages necessary. Voltages of several hundred volts are not uncommon. This requires special high-voltage electronics for control, and is a barrier to integration with standard electronics. Fortunately, miniature devices can handle these devices without dielectric breakdown (section 4.5.3).

Electrostatic actuators can be roughly categorized as linear or rotary [1]. Many types of the linear actuators will be covered in this chapter. The theory behind many of the rotary actuators is similar to their corresponding linear forms.

The main classes of rotary electrostatic actuators are [1]: (1) top-drive, (2) side-drive, and (3) wobble harmonic. These devices are often called micromotors. Despite their historical significant, rotary actuators will not covered.

## 11.1 Parallel-Plate Actuators

Parallel-plate actuators[2] are very well described by their name. These actuators can also be thought of as parallel-plate capacitors whose dielectric is a fluid. The simplest configuration consists of two plates, which are pulled together when a voltage is applied. Parallel-plate actuators are also often used in a differential configuration, which uses three plates. The analysis of differential parallel-plate systems is a simple extension of that presented here.

The main advantage of parallel-plate actuators in surface-micromachining is that they are very easy to make. Surface-micromachining is ideal for fabricating large flat plates separated by a thin region. Thus, relatively large capacitance, and so relatively large actuation forces, are readily available.

However, parallel-plate actuators suffer from a number of disadvantages. The most important disadvantage is that they are only easily implemented for actuation perpendicular to the substrate surface. Since a surface-micromachined system is likely going to be laid out across the wafer surface, this limits the amount of interaction these actuators can engage with their neighbouring components. Parallel-plate actuators can be used for in-plane actuation, but the implementation is more difficult.

The second disadvantage of parallel-plate actuators is the phenomenon of snap-in. As will be seen shortly, the force between the two plates varies non-linearly with the spacing between the plates. However, the restoring force typically varies linearly with the displacement. This limits the range of most parallel-plate actuators to about a third of the gap [2]. Larger deflections lead to a regime where the voltage increases much faster than the restoring force, and so the plates snap together. Without an insulating layer, this will lead to a short.

Snap-in is an issue of voltage control of the device. More elaborate control mechanisms can obtain larger displacements. For example, the use of an appropriately designed feedback capacitor circuit can extend actuator displacement to the full electrode gap [3–5]. Full displacement is also possible under current control of the device [5].

Despite the disadvantages with respect to surface-micromachined systems, they are often a good choice for actuating devices off-chip, since their displacement is perpendicular to the substrate surface. For example, large actuation distances are not necessary for the tips of scanning tunnelling microscopes [6].

An ideal parallel-plate actuator is like a parallel-plate capacitor. Between the plate, a uniform electric field exists [7, 8]. The capacitance between the plates is:

$$C = \frac{\epsilon A}{d} = \frac{\epsilon_0 \epsilon_r A}{d} \tag{11.1}$$

In equation 11.1, $C$ is the capacitance between two plates with area $A$ separated by a distance $d$. The dielectric between the two plates has a permittivity of $\epsilon$.

The fringing fields of the parallel-plate capacitor can usually be neglected. This is especially true in surface-micromachining, where building large flat plates is easy. However, the fringing fields can be included by approximating the edges as wedge capacitors [8].

The electrostatic field between the plates creates a force pulling the plates together. The magnitude of this force can be calculated using at least two methods: virtual-work [2, 8–10] and electrostatic pressure [9, 11]. With a formula for the capacitance already in hand, the method of virtual work is the simplest method.

The energy stored in a capacitor can be determined as follows:

$$W = \frac{CV^2}{2} \tag{11.2}$$

In equation 11.2, the stored energy, $W$, is a function of the capacitance, $C$, and the applied voltage, $V$. To calculate the force between the two plates, the method of virtual work starts with an expression for the work, in this case equation 11.2, and differentiates to find the force. Under the condition of constant applied voltage.

$$F = \mathrm{d}W/\mathrm{d}x \tag{11.3a}$$

$$F = \mathrm{d}CV^2/\mathrm{d}2x \tag{11.3b}$$

$$F = \frac{\mathrm{d}}{\mathrm{d}d}\frac{\epsilon_0\epsilon_r A}{2d} \tag{11.3c}$$

$$F = -\frac{\epsilon_0\epsilon_r A}{2d^2} \tag{11.3d}$$

In step 11.3c, the differential was changed to $d$ because that is the dimension we are changing to determine the force due to the plate separation. The negative sign of the final results indicates that the force is acting to reduce $d$.

The non-linear behaviour of $F$ for parallel-plate capacitors limits their use. As the plates move together, the electrostatic force increases dramatically. The plates, if they approach too closely, will overcome any restoring force and snap together, creating an electrical short.

Osterberg and Senturia [12] developed equations for modelling the deflection of several parallel-plate structures. The various structures, while all parallel plate devices, have different boundary conditions. The three structures investigated are singly supported beams, doubly supported beams, and circular diaphragms. These models were quite general, and included the effects of residual stress. However, the final results mixed in empirical fitting with analytical solutions.

The analytical model as initially developed by Osterberg and Senturia [12] models the one-dimensional stiffness of the beam or diaphragm as follows [12]:

$$K_{eff} = \frac{S}{L^2\left(1+\frac{2-2\cosh\left(\frac{kL}{2}\right)}{\left(\frac{kL}{2}\right)\sinh\left(\frac{kL}{2}\right)}\right)} \tag{11.4}$$

with,

$$k = \sqrt{\frac{12S}{B}} \tag{11.5}$$

$$S = \bar{\sigma} t g_0^3 \tag{11.6}$$

$$B = Et^3 g_0^3 \tag{11.7}$$

Above, $K_{eff}$ is the effective spring constant, since not all parts of the beam or diaphragm deflect equally. It makes use of non-dimensional parameters $S$ and $B$, with $S$ being the free parameter, and $B$ being the bending parameter. The other variables are the beam length or diaphragm radius, $L$, the uniform residual stress, $\bar{\sigma}$, the structure thickness, $t$, and the initial electrode gap, $g_0$. Note that the variable, $E$, refers to the biaxial elasticity except in the case of thin cantilevers (section 5.1).

The pull-in voltage can be calculated using the following equation [12]:

$$V_{\mathrm{PI}} = \sqrt{\frac{8K_{eff}g_0^3}{27\epsilon\left(1+\gamma_{3n}\frac{g_0}{w}\right)}} \tag{11.8}$$

However, Osterberg and Senturia [12] decided to add several empirical terms, and did not use the above equation. The empirical terms were then fitted to simulated data. Details of this process can be found in their paper [12].

### 11.1.1 Beams

As beams are an important category, their behaviour under electrostatic actuation deserves more careful attention. The key addition is the inclusion of fringe fields in the electrostatic force. For example, neglecting fringing fields can lead to errors of nearly 25% in the capacitance of beams with widths of 30 $\mu$m, thicknesses of 2 $\mu$m, and electrode spacings of 2 $\mu$m [13].

This electrostatic force, per unit length, can be determined as follows [13]:

$$p(x) = \frac{\epsilon w V^2}{2h(x)^2}\left(a + \frac{h(x)}{b}\right) \tag{11.9}$$

In equation 11.9, the force per unit length, $p(x)$, is determined from the permittivity, $\epsilon$, the beam width, $w$, the applied voltage, $V$, the gap between the ground plane and the beam, $h(x)$, and two empirical fitting constants[3], $a$ and $b$.

The capacitance of the beam can also be determined [13]:

$$C(x) = \frac{\epsilon w}{h(x)}\left(c + \frac{2h(x)}{d} + \frac{2h(x)}{d}\ln\left(\frac{d}{h(x)}\right)\right) \tag{11.10}$$

Above, the capacitance uses two new fitting parameters, $c$ and $d$. For the same beam as just discussed, these parameters were found to be $c = 1.004$ and $d = 75.4\ \mu$m.

If the beam does not have a ground plane, and uses the substrate as the second electrode, then the isolation layer must be included. This results in the following equation [13]:

$$h(x) = h_0(x) + \frac{h_{\text{isolation}}}{\epsilon_r} \tag{11.11}$$

Above, the effective gap between the electrodes, $h(x)$, is a function of the gap between the top electrode and the isolation layer, $h_0(x)$, the isolation layer thickness, $h_{\text{isolation}}$, and the relative permittivity of the isolation layer, $\epsilon_r$.

Unfortunately, equations 11.9 and 11.10 are not amenable to analytical solution. Some further approximation[4] or numerical analysis is required to extract the beam displacement.

### 11.1.2 In-plane Parallel-plate Actuators

Parallel-plate actuators can also be designed to operate parallel to the substrate surface (figure 11.1). Because surface-micromachined devices are thin, it is often necessary to include multiple parallel-plate structures to obtain enough force[5].

The force for each electrode pair can be estimated using equation 11.3d, although fringe fields will play an important role in these actuators.

Note that using multiple electrodes results in a parasitic capactive gap, and thus a parasitic force [2]. Unfortunately, there is no analytical solution for the maximum displacement before pull-in, although numerical solutions are still possible [2]. The actuator can thus be modeled as two variable capacitors [2]:

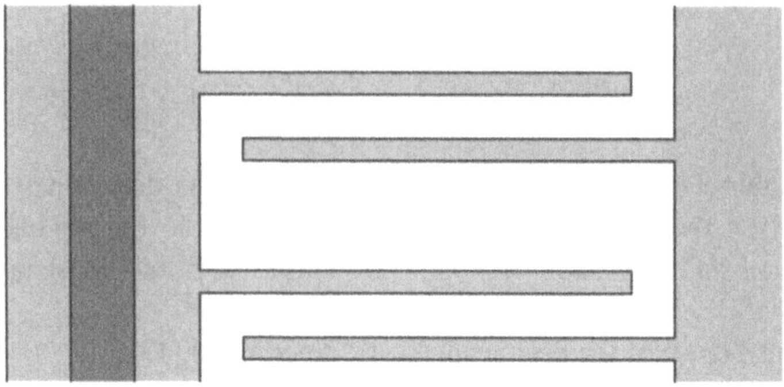

*Figure 11.1.* Illustration of an in-plane parallel-plate actuator. The stator is on the left while the rotor is on the right. Motion of the rotor is upwards. Anchor is indicated by dark gray.

$$C(d) = C_f + N\epsilon A\left(\frac{1}{g_0 - d} + \frac{1}{g_1 + d}\right) \tag{11.12}$$

In equation 11.12, $C(d)$ is the capacitance between the stator and the rotor as a function of the rotor's displacement, $C_f$ is the fixed or displacement independent capacitance, $N$ is the number of electrode pairs, $\epsilon$ is the permittivity, $A$ is the area of each capacitive plate, $g_0$ is the short electrode gap, and $g_1$ is the large or parasitic electrode gap, and $d$ is the actuator displacement. Note that $A = Lt$, with $L$ being the electrode overlap length, and $t$ being the thickness.

With the capacitance known, the force can be easily determined [2]:

$$F(d) = \frac{N\epsilon A}{2}\left(\frac{1}{(g_0 - d)^2} - \frac{1}{(g_1 + d)^2}\right)V^2 \tag{11.13}$$

In addition to the above, Horsley *et al* [2] developed an expansion for applied AC voltages. They used this with a mechanical model to predict deflection amplitudes under applied sinusoidal voltages.

Stoppers can be placed to prevent the electrodes from shorting [2]. These stoppers can be placed to prevent motions larger than the maximum displacement before snap-in.

### 11.1.3 Digital Micromirror Devices

Digital micromirror devices are a commercially important example of parallel-plate actuation. The mirror is supported on a central post. Electrodes exist below two corners of the mirror, forming parallel-plate actuators. When a voltage is applied to one of the bottom electrodes, the force of attraction pulls the mirror over.

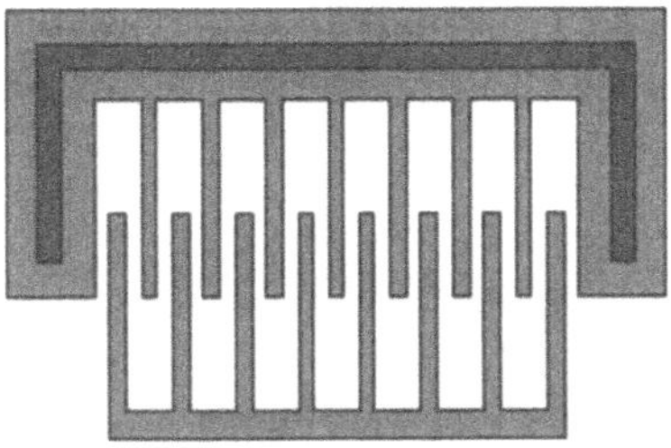

*Figure 11.2.* Illustration of a linear electrostatic comb actuator. In this example, the top comb is anchored to the substrate, as indicated by the dark region.

## 11.2 Interdigitated Comb Actuators

Comb-drive actuators consist of two interdigitated comb structures (figure 11.2). Applying a voltage between the two combs results in an attractive force pulling the combs together [14–16]. If properly designed and fabricated, the interdigitated comb structure will be symmetric, leading to no transverse forces. An SEM of a fabricated interdigitated comb resonator is shown in figure 3.10.

The force between the combs is independent of the position between the combs [14–16]. In many applications, this is an advantage[6]. This, combined with their longer travel, make them preferred for long travel electrostatic actuators.

However, interdigitated comb actuators produce less force per area [2]. Comparing equivalent parallel-plate and comb configuration[7], the force in a parallel-plate actuator is $l/g$ times greater than then comb actuator, with $l$ being the electrode length and $g$ being the electrode gap. Since electrode lengths are typically over an order of magnitude greater than the gap [2], parallel-plate actuators typically provide over a magnitude more force for the same area[8].

In a typical configuration, the moving comb of the actuator[9] is connected to a suspension [14–16] (section 10.3). This allows the combs to move axially, but also prevent transverse motion. Since electrostatic forces increase with decreasing gap, the equilibrium point balancing the transverse forces in the interdigitated comb is unstable. Thus, a small shift away from the centre can lead to the comb snapping together and shorting. Since no fabrication process is perfect, additional restoring forces must be provided, and so suspensions are preferred.

### 11.2.1 Axial Displacement

Looking at a single finger of the rotor, the finger is surrounded on both sides by a finger from the stator. An estimate of the capacitance can be obtained, if fringing fields are neglected, using the following equation [16]:

$$C \doteq \frac{\epsilon h L_{\text{overlap}}}{g} \tag{11.14}$$

In equation 11.14, the capacitance, $C$, is calculated based on the permittivity, $\epsilon$, film thickness, $h$, the overlap of the finger, $L_{\text{overlap}}$, and gap between the fingers, $g$. It is important to note that the capacitance varies linearly with the overlap.

The total capacitance between the rotor and the stator is simply $C' \doteq CN$, where $N$ is the number of fingers in the rotor.

The force between the two combs can be calculated using the method of virtual work and assuming constant voltage control [9, 10, 16]:

$$F = \frac{\mathrm{d}C}{\mathrm{d}L_{\text{overlap}}} V^2 \tag{11.15}$$

$$\doteq \frac{\epsilon h V^2}{g} \tag{11.16}$$

If the rotor is connected to a suspension, then the displacement of the rotor and the suspension's shuttle can be easily calculated using Hooke's law:

$$\Delta y = \frac{\epsilon h V^2}{k_y g} \tag{11.17}$$

Unfortunately, the prediction provided by equation 11.17 suffers from the difficulty in determining $k_y$ (section 10.3.1). This makes open-loop control of the displacement difficult without calibrating for $k_y$. However, the position can be easily determining by measuring the capacitance of the comb-drive, and so closed-loop control is possible [16].

In addition to knowing the static displacement, the dynamic displacement is also important. If a lumped parameter approximation is used [16], then the following equation describes the shuttle's motion:

$$M_{eq}\frac{\mathrm{d}^2 y}{\mathrm{d}t^2} + b\frac{\mathrm{d}y}{\mathrm{d}t} + k_y y = \frac{\epsilon h}{g} V(t)^2 \tag{11.18}$$

In equation, 11.18, $M_{eq}$ is the equivalent mass of the comb drive and the shuttle, $b$ is the eqivalent viscous drag, and $V(t)$ is the applied voltage. If the static deflection profile is used as an approximation for the fundamental mode shape, then the resonance frequency can be obtained using Rayleigh's quotient [16]:

$$\omega_0 = \sqrt{\frac{k_y}{M_{eq}}} \tag{11.19}$$

The resonance frequency is ideally independent of the DC and AC components of $V(t)$.

A more detailed analysis of the frequency-dependent response was provided by Kuehnel [17]. He provided a lumped-parameter model that can be used to plot the deflection as a function of frequency, as well as determine the quality factor.

## 11.2.2 Transverse Displacement

There are also forces pulling the fingers together in the transverse direction. Ideally, these forces cancel, but the equilibrium point is unstable. The transverse electrostatic force can be found as follows [16]:

$$F = \frac{\epsilon h L_{\text{overlap}}}{2(g-x)^2}V^2 - \frac{\epsilon h L_{\text{overlap}}}{2(g+x)^2}V^2 \tag{11.20}$$

As can be seen from the force balance, the forces on either side of the combs' fingers balance as long as the fingers are centered in the gap. However, if the finger are perturbed, then the forces on one side will increase, further pulling the finger out of alignment. Hence, the need for suspensions that resist the lateral motion and make the equilibrium stable.

The maximum operating voltage can be found by equating $\mathrm{d}F/\mathrm{d}x$ for the suspension and the electrostatic force. This leads to an expression for the maximum stable operating voltage [16]:

$$V_{\text{SI}}^2 < \frac{2\epsilon h L_{\text{overlap}} k_x}{g^3} \tag{11.21}$$

In equation 11.21, $V_{\text{SI}}$ is the side-instability voltage. If the voltage exceeds the limit imposed by the right hand side of equation 11.21, the comb-drive will become unstable, and the rotor will pull to one side until it comes into contact with the stator. When combined with equation 11.17, an expression for the side-instability voltage as a function of the displacement can be obtained. Under the assumption $k_x \gg k_y$, the expression simplifies to [16]:

$$\Delta y_{\text{SI}} = g\sqrt{\frac{k_x}{2k_y}} \tag{11.22}$$

Equation 11.22 is useful because it clearly relates the actuator's maximum travel to design parameters. However, other sources of instability exist, such as compliance in the fingers themselves.

While an interdigitated comb can be used to create a force, the complete design for a comb drive should include the suspension. The actuator's force-

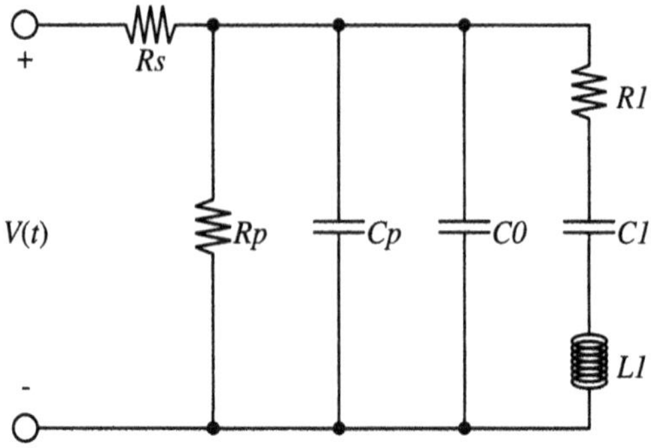


*Figure 11.3.* Circuit model, including parasitics, of comb-drive actuator.

deflection curve and maximum deflection depend on both the design of the comb and the suspension.

### 11.2.3 Electrical Characteristics

The interdigitated combs are essentially a variable capacitor. Combining equation 11.18 with equations relating charge, voltage, electrostatic force, and displacement [16], then the electrical admittance can be determined to be:

$$Y(j\omega) = j\omega C_0 + \frac{j\omega C_1}{H(j\omega)} \tag{11.23}$$

In equation 11.23, $Y$ is the admittance as a function of frequency, $C_0$ is the static capacitance, $C_1$ is the motional capacitance, and $Q$ is the quality factor. These parameters can be calculated using the following equations:

$$C_0 = \frac{2\epsilon h L_{\text{overlap}}}{g} \tag{11.24a}$$

$$C_1 = \frac{4\epsilon^2 h^2 \mu_0^2}{k_y g^2} \tag{11.24b}$$

$$H(j\omega) = \left(\frac{j\omega}{\omega_0}\right)^2 + \frac{j\omega}{Q\omega_0} + 1 \tag{11.24c}$$

The circuit diagram for the comb drive, including parasitics, is shown in figure 11.3. Additional parameters are used in figure 11.3. This model includes three parasitics: $R_s$, the series resistance; $R_p$, the parallel resistance; and $C_p$, the parasitic capacitance. Additionally, the values for $L_1$ and $R_1$ need to be calculated.

$$L_1 = \frac{1}{\omega^2 C_1} \tag{11.25a}$$

$$R_1 = \frac{1}{Q\sqrt{L_1/C_1}} \tag{11.25b}$$

Legtenberg *et al* [16] measured the admittance of their comb-drive actuators. They found the series paraistic resistance, $R_s$, to be negligible, and the parallel resistance, $R_p$, to be 17.6 G$\Omega$. The parasitic capacitance cannot be seperated from $C_0$ (figure 11.3), but they measured the combined capacitance to be 414.5 fF. Additionally, they measured $C_1 = 145.6$ fF, $L_1 = 68.4$ kH, and $R_1 = 205$ M$\Omega$. Their comb actuator thus had a resonant frequency of approximately 1.60 kHz and a quality factor of 3.34.

Operating comb-drive actuators in atmosphere leads to low quality factors due to air damping. Quality factors less than 100 [18] should be expected. However, to be useable in mechanical filters, quality factors in the range of 100 to 10000 are required [18]. This is possible if the comb-drive is operating in vacuum, but then the system must be hermetically sealed[10].

For example, Lin *et al* [18] constructed and operated comb-drives with quality factors of 2200. The operating pressure was between 200 and 300 mTorr. Stronger vacuums can increase the quality factor further, with quality factors reaching 50000[11] [18] to 80000 [19].

The importance of comb-drives in combination with suspensions should not be underestimated. They have many applications, including measuring acceleration and signal filtering [19].

### 11.2.4 Levitation

With a ground plane, it is important to note that comb-drive actuators may levitate [15]. Interactions between the moveable comb, which should also be at ground potential, the immobile comb, and the ground plane are responsible.

The voltage difference between the immobile comb and other components creates surface charges on the various components. Since the mobile comb and the ground plane are at the same potential, like charges accumulate on their surfaces. This leads to the mobile comb and the ground plane repelling each other. Hence, the mobile comb may levitate. In practice, this levitation is offset by the stiffness, $k_z$, of the suspension.

## 11.3 Curved Electrode Actuators

Curved electrode actuators, also known as zipping electrode actuators, are designed to address an important limitation of electrostatic actuators. Namely, large actuator displacements and large actuator forces have contrary requirements [20]. Large displacement actuators, for example comb-drives, have dis-

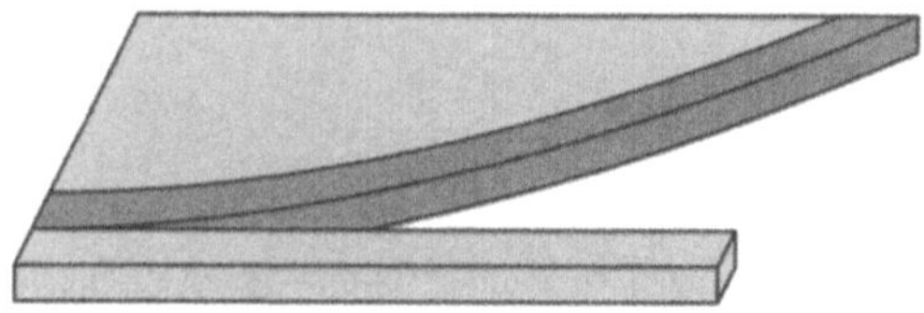

*Figure 11.4.* Illustration of a curved electrode actuator. The dark grey region is insulating. The light gray region is conducting.

placements perpendicular to the major field lines. To create a large force, the actuator must move with the field lines, such as in a parallel-plate actuator. However, this requires the actuator move in the direction of the small gaps and limits the amount of displacement possible. Curved electrode actuators, or zipping actuators, try to obtain both large displacements and large forces.

An example of a curved electrode actuator is shown in figure 11.4. The gap between the electrodes is small near the base, but large near the free end. Under an applied voltage, electrostatic forces pull in the beam in towards the curved electrode. The actuation is parallel to the wafer surface. This means that the curved surfaces of the device can be created through photolithography.

Care must be taken to prevent shorting between the electrodes. Although figure 11.4 shows an insulating barrier between the electrodes, the insulating barrier is not necessary. The insulating layer can be replaced with electrically isolated bumpers. Curved electrode actuators can thus be fabricated using all conducting surface-micromachining processes.

Legtenberg *et al* [20] investigated curved electrode actuators where the curve was described by simply polynomial expressions of the form:

$$s(x) = \delta_{\max}\left(\frac{x}{L}\right)^n, \quad n \geq 0 \tag{11.26}$$

Above, $s$ describes the shape of the curved electrode, $\delta_{\max}$ is the maximum gap distance, $x$ is the position along the x-axis, $L$ is the maximum length of the beam, and $n$ is the polynomial order.

The polynomial order, $n$, has an important effect on the actuator behaviour [20]. Increasing the polynomial order decreases the pull-in voltage for the actuator without significantly changing the maximum displacement prior to pull-in. Once pull-in occurs, the beam will snap to the curved electrode, making the insulating layer necessary.

If the gap spacing is small compared to the beam's length, than the electric field is mostly perpendicular to the beam and a parallel-plate approximation can be used. If fringing fields are ignored, this leads to a one-dimensional model. With a constant voltage applied between the two electrodes, a force is developed the electrodes. This can be placed as a load into the beam-bending equation (section 5.1).

$$EI\frac{d^4w(x)}{dx^4} = p(x) \tag{11.27}$$

$$EI\frac{d^4w(x)}{dx^4} = \frac{1}{2}\frac{\epsilon_0 hV^2}{\left(\frac{d}{\epsilon_r} + s(x) - w(x)\right)^2} \tag{11.28}$$

In the above equation, the variables have their standard definitions. The beam's load, $p(x)$, has been expanded based on a parallel plate approximation for the electric field, and hence electrostatic pressure, in the electrode gap. Unfortunately, the above equation does not have a known closed-form solution [20]. Instead, a Rayleight-Ritz method is used to find an approximation. Details of the derivation can be found in Legtenberg *et al* [20].

For curved electrodes with $n > 2$, the beam's deflection is constrained by the curved electrode before pull-in occurs [20]. This situation was also investigated by Legtenberg *et al* [20], with the addition of an end-load applied to the beam. In this model, the beam is in partial contact with the curved electrode. At some distance along the beam, the beam separates from the curved electrode. Expressions for the displacement of the beam end were provided.

The curved electrode actuators investigated by Legtenberg *et al* [20] were capable of producing forces of a few micro-Newtons. The force can be increased by decreasing the initial insulator thickness[12] or increasing the insulator's dielectric constant.

Operation of the curved electrode actuators used voltages up to 120 volts [20]. Like other electrostatic actuators discussed, curved electrode actuators require significant voltage for operation, although very little current.

Out-of-plane curved electrode actuators can also be fabricated using surface micromachining [21, 22]. However, the out-of-plane actuators make use of non-uniform residual stresses. The top electrode was fabricated flat as would be expected in surface-micromachining, but then relaxed after release to a curved shape due to non-uniform residual stresses.

The out-of-plane curved electrode actuator developed by Chen *et al* [21] had an operating voltage of 20 volts and a switching time of 600 $\mu$s. The tip of the curved electrode was 300 $\mu m$ above the substrate surface.

## 11.4 Scratch-drive Actuators

Unfortunately, all of the electrostatic actuators discussed so far suffer from limited range of travel. Scratch-drive actuators (SDAs) move a small distance with each step, but steps can be accumulated to get larger travel. SDAs are

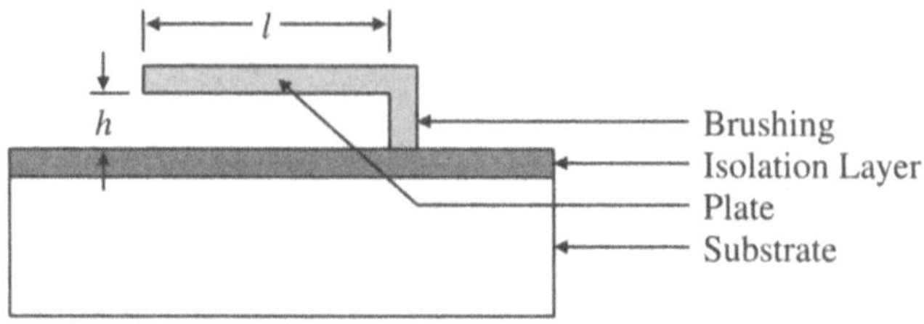


*Figure 11.5.* Illustration of the components of a scratch-drive actuator.

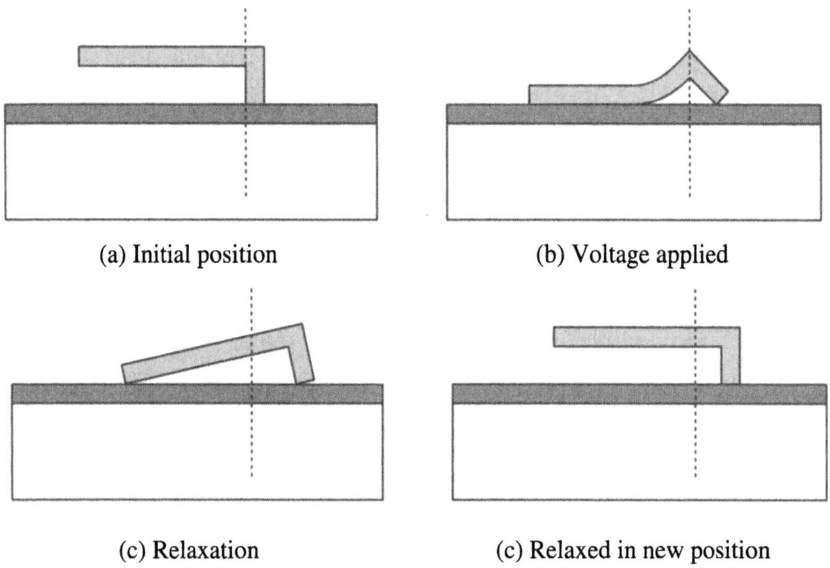


*Figure 11.6.* A single step of a scratch-drive actuator [23, 24].

thus suited as long-travel linear actuators, although they can be used to create rotary actuators as well [23].

The structure of SDAs is illustrated in figure 11.5. Not shown are electrical connections to both the substrate and the plate. The plate and the substrate, both of which must be conductive, form the two electrical terminals for these devices. Shorting is prevented by the isolation layer [24].

At rest, there is no voltage difference between the plate and the substrate, and the SDA is configured as shown in figure 11.6a[13]. A step is initiated by applying a voltage, which creates electrostatic attraction between the plate and the brushing. This pulls the plate down into contact with the isolation layer. However, one end of the plate is supported by the brushing, and contact is not made (figure 11.6b). As the plate is bent, the tip of the brushing shifts.

In addition to electrostatic energy, mechanical energy is stored in the elastic bending of the plate and brushing. When the voltage is removed, the plate and brushing return to their original shape (figure 11.6c) [23, 24]. While the brushing returns to its original shape, it also pulls itself a small distance along the isolation layer.

While each step is small, an SDA can move further by taking repeated steps [23, 24]. This displacement is not unlimited, as the forward motion is accompanied by beam bending.

The length of each step, $\Delta x$ is controlled by the peak voltage, $V$, the brushing height, $h$, and the plate length, $l$. For a given operating condition, there is an ideal plate length. Plates which are too short or too long will prevent stepping. For a brushing height of 1 $\mu$m, the step length is in the range of 10-30 nm, and, for a height of 2 $\mu$m, the step length is in the range of 40-80 nm [23].

The step length can be predicted if the contact length, $l'$, is known. The contact length is a measure of how much of the plate is in contact with the isolation at the peak voltage. From geometric considerations [23]:

$$\Delta x = \frac{h^2}{2(l - l')} \tag{11.29}$$

The variable $l'$ can be determined by a combined electrostatic-mechanical analysis of the structure. The step length therefore depends on the plate length, brushing height, and peak voltage [24]. The devices contructed by Akiyama *et al* [24] had measured step lengths in the range of 0.1 $\mu$m.

The actuator speed is also controllable. Each step occurs in response to an electrical pulse. Controlling the pulse frequency, and therefore the step frequency, controls the actuator velocity. The velocity can be calculated simply from the pulse frequency, and the step length [23, 24].

$$v = \Delta x f \tag{11.30}$$

Scratch-drive actuators are also interesting because of the relatively large forces they create. The electrostatic attraction between the plate and the substrate can be fairly large when compared to other electrostatic actuators[14].

The SDA, as illustrated in figure 11.5, is incomplete. Electrical connections to the plate, which must be mobile, are necessary. These electrical connections are typically made through beams, oriented perpendicular to the direction of motion, which are connected railings [23]. The beams acts as torsional springs (section 5.1), and create a threshold voltage for actuation. Below the threshold voltage, the stiffness of the beams prevents the plate from coming into contact with the isolation layer.

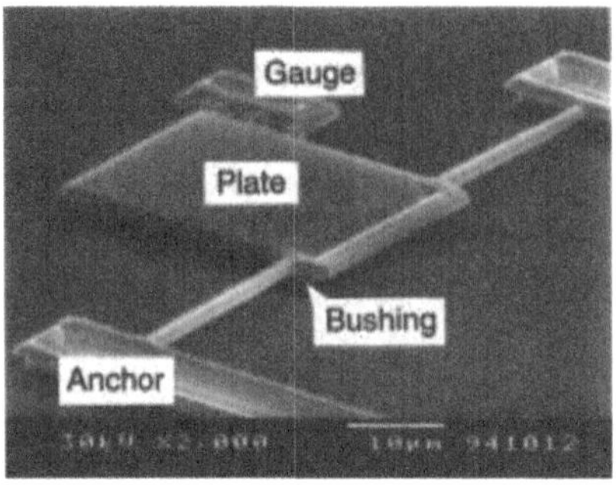


*Figure 11.7.* SEM of a scratch drive actuator [24]. Copyright © 1997 IEEE.

Akiyama *et al* [24] fabricated a wide array of SDAs. They had widths ranging from 50 $\mu$m to 100 $\mu$m and lengths ranging from 35 $\mu$m to 80 $\mu$m. These actuators produced forces in the range of 10 $\mu$N to 63 $\mu$N. The operating voltage was 150 volts.

## Notes

1 Between two bodies, this force is always attractive, no matter the field polarity. If there is a third body, then the net force may be negative.

2 These actuators are also known as squeeze-film actuators.

3 For example, $a = 1.006$ and $b = 31.6$ $\mu$m for beam's with widths of 30 $\mu$m and thicknesss of 2 $\mu$m.

4 For example, by assuming a form for $h(x)$.

5 These resemble interdigitated comb actuators described later this chapter. The key difference is that the electrodes are not in the middle of the gap.

6 For example, xy-stages [16].

7 Both structures have the same gap and electrode length.

8 Typical gap lengths are in the range of 1 $\mu$m to 10 $\mu$m. Typical electrode lengths are in the range of 10 $\mu$m to 100 $\mu$m.

9 The moving comb of the actuator is known as the rotor. The fixed comb is the stator.

10 Depending on technology, this can be done at the device level, the wafer level, or the packaging level.

11 This corresponds to a vacuum of $10^{-7}$ Torr.

12 Or initial electrode gap, if no insulator is being used.

13 The brushing need not be in contact with the isolation layer.

14 Akiyama *et al* reported forces of approximately $10^{-2}$ N.

## References

[1] P. Dario, R. Valleggi, M.C. Carrozza, M.C. Montesi, and M. Cocco. Microactuators for microrobots: a critical survey. *Journal of Micromechanics and Microengineering*, 2(3):141–157, 1992.

[2] D.A. Horsley, N. Wongkomet, R. Horowitz, and A.P. Pisano. Precision positioning using a microfabricated electrostatic actuator. *IEEE Transactions on Magnetics*, 35(2):993–999, 1999.

[3] J.I. Seeger and S.B. Crary. Stabilization of electrostatically actuated mechanical devices. In *Proceedings of 1997 International Conference on Solid-State Sensors and Actuators*, pages 1133–1136. IEEE, 1997. 1997 Jan 16-19.

[4] E.K. Chan and R.W. Dutton. Electrostatic micromechanical actuator with extended range of travel. *Journal of Microelectromechanical Systems*, 9(3):321–328, 2000.

[5] R. Nadal-Guardia, A. Dehe, R. Aigner, and L.M. Castaner. Current drive methods to extend the range of travel of electrostatic microactuators beyond the voltage pull-in point. *Journal of Microelectromechanical Systems*, 11(3):255–263, 2002.

[6] T.W. Kenny, S.B. Waltman, J.K. Reynolds, and W.J. Kaiser. Micromachined silicon tunnel sensor for motion detection. *Applied Physics Letters*, 58(1):100–102, 1991.

[7] R. Serway. *Physics for Scientists and Engineers with Modern Physics*. Saunders College Publishing, Philadelphia, 3 edition, 1990.

[8] S. Zappe, M. Baltzer, T. Kraus, and E. Obermeier. Electrostatically driven linear microactuators: Fe analysis and fabrication. *Journal of Micromechanics and Microengineering*, 7(3):204–209, 1997.

[9] D.J. Griffiths. *Introduction to Electrodynamics*. Prentice Hall, Englewood Cliffs, New Jersey, 2 edition, 1989.

[10] W.A. Johnson and L.K. Warne. Electrophysics of micromechanical comb actuators. *Journal of Microelectromechanical Systems*, 4(1):49–59, 1995.

[11] R.W. Johnstone and M. Parameswaran. Self-assembly of surface-micromachined structures using electrostatic attraction. *Proceedings of the SPIE*, 4561:66–76, 2001.

[12] P.M. Osterberg and S.D. Senturia. M-test: A testchip for mems material property measurement using electrostatically actuated test structures. *Journal of Microelectromechanical Systems*, 6(2):107–118, 1997.

[13] E.K. Chan, K. Garikipati, and R.W. Dutton. Characterization of contact electromechanics through capacitance-voltage measurements and simulations. *Journal of Microelectromechanical Systems*, 8(2):208–217, 1999.

[14] W.C. Tang, T.-C.H. Nguyen, and R.T. Howe. Laterally driven polysilicon resonant micromstructures. *Sensors and Actuators*, 20(1-2):25–32, 1989.

[15] W.C. Tang. *Electrostatic comb drive for resonant sensor and actuator applications*. PhD thesis, University of California at Berkeley, Berkeley, California, 1990.

[16] R. Legtenberg, A.W. Groeneveld, and M. Elwenspoek. Comb-drive actuators for large displacements. *Journal of Micromechanics and Microengineering*, 6(3):320–329, 1996.

[17] W. Kuehnel. Modelling of the mechanical behaviour of a differential capacitor acceleration sensor. *saa*, 48(2):101–108, 1995.

[18] L. Lin, R.T. Howe, and A.P. Pisano. Microelectromechanical filters for signal processing. *jmems*, 7(3):286–294, 1998.

[19] C.T.-C. Nguyen and R.T. Howe. An integrated cmos micromechanical resonator high-q oscillator. *IEEE Journal of Solid-State Circuits*, 34(4):440–455, 1999.

[20] R. Legtenberg, J. Gilbert, and S.D. Senturia. Electrostatic curved electrode actuators. *Journal of Microelectromechanical Systems*, 6(3):257–265, 1997.

[21] R.T. Chen, H. Nguyen, and M.C. Wu. A high-speed low-voltage stress-induced micromachined 2 x 2 optical switch. *IEEE Photonics Technology Letters*, 11(11):1396–1398, 1999.

[22] A.K. Chinthakindi, D. Bhusari, B.P. Dusch, J. Musolf, B.A. Willemsen, E. Prophet, M. Roberson, and P.A. Kohl. Electrostatic actuators with intrinsic stress gradient - i. materials and structures. *Journal of the Electrochemical Society*, 149(8):H139–H145, 2002.

[23] T. Akiyama and K. Shono. Controlled stepwise motion in polysilicon microstructures. *Journal of Microelectromechanical Systems*, 2(3):106–110, 1993.

[24] T. Akiyama, D. Collard, and H. Fujita. Scratch drive actuator with mechanical links for self-assembly of three-dimensional mems. *Journal of Microelectromechanical Systems*, 6(1):10–17, 1997.

Chapter 12

# ELECTRO-THERMAL-COMPLIANT ACTUATORS

In electro-thermal-compliant (ETC) actuators, a voltage is applied to the electrical terminals of the device. The potential difference induces currents to flow throughout the device, which cause the temperature to rise due to Joule heating. ETC actuators use the resulting thermal expansion to create motion. Typically, they also have some method of amplifying these small displacements into useable motion.

The design of ETC actuators depends on the distribution of conductivity and stiffness. The various actuators discussed in this chapter can be identified as ETC actuators because of two features:

1 Under an applied voltage, the device heats significantly. Further, the heating is not uniform. Temperature differences occur, either between different parts of the device or between the device and the substrate.

2 The actuators have a mechanism to amplify the small changes in geometry due to thermal expansion into a larger motion. Thermal expansion creates large forces, but small deflections [1], so geometric advantage will usually be necessary.

In general, ETC actuators provide larger forces[1] than electrostatic actuators, but they also consume more power [2]. Another important advantage of ETC actuators is that they use much smaller voltages [1].

While this chapter explicitly discusses electro-thermal-compliant actuators, many of these techniques will work with other sources of heat besides Joule heating [1]. For example, increased temperature could be caused by solar power, laser activation, or exothermic reactions [1].

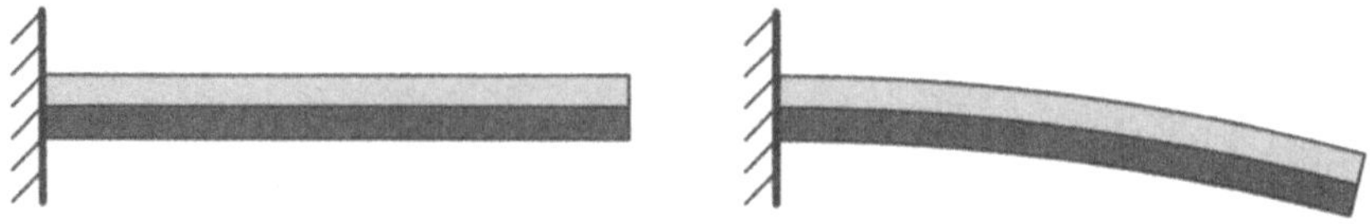

(a) room temperature (b) elevated temperature

*Figure 12.1.* Illustration of a thermal bimorph actuator. The light grey material has a greater coefficient of thermal expansion than the dark grey material.

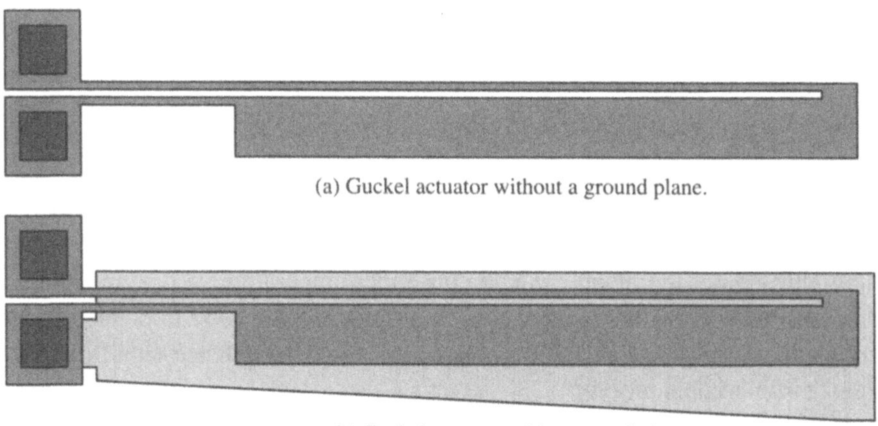

(a) Guckel actuator without a ground plane.

(b) Guckel actuator with a ground plane.

*Figure 12.2.* Layouts for two different Guckel actuator. Light gray is ground plane, medium gray is structural material, and dark gray is anchor.

## 12.1 Thermal Bimorph Actuators

Thermal bimorph actuators are not particularly useful in surface-micromachining. However, they do help to illustrate the operation of Guckel actuators, which are very important devices in surface-micromachining.

Thermal bimorph actuators are created by placing two materials with different coefficients of thermal expansion (CTE) together. Typically, the two materials are shaped into a cantilever (figure 12.1). When the temperature rises, the materials do not expand equally because of their differing CTEs. This creates a net bending moment in the cantilever, forcing the tip to move.

## 12.2 Guckel Actuators

In Guckel actuators[2], a pseudo-bimorph is created. Instead of using two dissimilar materials, the actuator consists of two beams made of the same material, but with different cross-sectional areas [1, 3]. Because both beams are made from the same material, Guckel actuators can be easily created in any conducting structural layer of most micromachining process [1].

The heating in Guckel actuators occurs when a current is passed through the device [1,3,4]. Because one beam has a narrower cross-section, it experiences higher current densities, and so experiences more heating. When steady-state is reached, the narrower arm will be hotter than the wider arm. This gives rise to the name hot-arm and cold-arm for the narrow and wide arms respectively [1,3,4]. An additional piece of nomenclature, the thin section of the cold-arm is called the flexure.

Similar to a thermal bimorph, the two parts of the cantilever experience different amounts of thermal expansion. However, in Guckel actuators, the different amounts of thermal expansion exist because of the differential heating. The efficiency of Guckel actuators is improved by increasing the temperature difference between the hot and cold arms. For example, etching a trench under the cold-arm to increase its thermal isolation [1,5].

The tips of Guckel actuators follow arcs [1,4]. However, the radius of curvature is large enough that it can often be neglected, and if the actuator's connections are flexible segments, this is not usually a problem. However, when Guckel actuators are connected together using rigid links that prevent the necessary unwanted deflections, the desired deflection will also be significantly reduced.

To make the analysis of Guckel actuators tractable, a one dimensional model can be used. The actuator is broken up into three segments. If the anchored end of the hot-arm is used as the origin, then the hot-arm goes from zero to $l$, the cold-arm goes from $l$ to $l + l_c$, and the flexure goes from $l + l_c$ to $2l$.

The effect of temperature on the actuator's resistivity, $\rho$, must be included. The simplest model is to model the resistivity as having a linear relationship with temperature:

$$\rho(T) = \rho_0[1 + \alpha_e(T - T_\infty)] \tag{12.1}$$

Above, $\rho_0$ is the resistivity at the ambient temperature, $\alpha_e$ is the temperature coefficient of resistivity, $T$ is the local temperature, and $T_\infty$ is the ambient temperature. To avoid stresses caused by thermal mismatch between the actuator and the substrate, the substrate should be at the ambient temperature.

Using the temperature dependent resistivity, one can now write an equation to model the temperature profile of the actuator [5,6]:

$$-k_p \frac{\mathrm{d}^2 T}{\mathrm{d}x^2} + J^2 \rho = \frac{S}{h} \frac{T - T_\infty}{R_T} \tag{12.2}$$

with,

$$S = \frac{t}{w}\left(\frac{h}{t}+1\right)+1 \tag{12.3}$$

In equation 12.2, $k_p$ is the thermal conductivity, $S$ is a shape factor[3], and $R_T$ is the thermal resistance. For the shape factor, $t$ is the actuator thickness, $w$ is the width of the beam, $h$ is the height of the beam above the substrate surface. Note that equation 12.2 does not account for heat loss due to radiation. It has been shown that radiative heat transfer does not play a large role [5], especially in surface-micromachined devices[4].

The solution of the above equation provides the following form for the temperature as a function of $x$:

$$T(x) = T_\infty + \frac{J^2\rho_0}{k_p m^2} + c_1 e^{mx} + c_2 e^{-mx} \tag{12.4}$$

$$m = \frac{S}{k_p t R_T} - \frac{J^2\rho_0\alpha_t}{k_p} \tag{12.5}$$

In equation 12.4, the temperature depends on variables already introduced, plus a new variable, $m$. However, the integration constants in equation 12.4, $c_1$ and $c_2$, must be solved independently for each of the three segments of the actuator. Thus, in reality there is $c_{1,hot}$, $c_{2,hot}$, $c_{1,cold}$, $c_{2,cold}$, $c_{1,flex}$, and $c_{2,flex}$. When solving, remember to use the different beam widths when calculating $J$ and $S$.

If the anchors are assumed to be at the same temperature as the substrate, and thus at the ambient temperature, and noting that the temperature must be continuous, then the six integration constants can be readily determined [5].

Once the temperature profile has been determined, one can integrate along the temperature dependent resistivity along the actuator to determine the total actuator resistance. This can be simplified to use the average temperature of each segment [5]:

$$\begin{aligned} R = {} & \frac{L_h\rho}{w_h t}\left(1+\alpha_e(\overline{T_h}-T_\infty)\right) + \frac{L_c\rho}{w_c t}\left(1+\alpha_e(\overline{T_h}-T_\infty)\right) \\ & + \frac{L_f\rho}{w_f t}\left(1+\alpha_e(\overline{T_h}-T_\infty)\right) \end{aligned} \tag{12.6}$$

Above, the variables $L_h$, $L_c$, and $L_f$ refer to the lengths of the hot-arm, cold-arm, and flexure respectively. Similarly, the other variables refer to the widths and average temperatures of the various segments.

An estimate of the deflection of the actuator can be obtained from the following equation [3]:

$$d = \frac{\alpha(\overline{T_h} - \overline{T_c})L^2}{g\left(0.7707 + 0.3812\frac{w^2}{g^2}\right)} \quad (12.7)$$

In equation 12.7, the displacement, $d$, depends on the coefficient of thermal expansion[5], $\alpha$, the average temperature in the hot and cold arms, $\overline{T_h}$ and $\overline{T_c}$, the gap between the arms, $g$, and the width of the cold-arm and the flexure, $w$.

As an example, Comtois and Bright [1] designed and fabricated Guckel actuators that were 220 $\mu$m long from 2 $\mu$m polysilicon that were capable of up to 16 $\mu$m of deflection. While consuming 10.8 mW, it required 4.4 $\mu$N of force to reduce the deflection to 8 $\mu$m. It is important to note that the power was delivered at 1.94 V, well within the range of standard CMOS circuitry. The actuators could still provide maximum deflection at up to 300 Hz. The actuators can be operated at higher frequencies, but the deflection amplitude is reduced.

Comtois *et al* [4] presented alternate designs that were 250 $\mu$m long. These devices were capable of producing an unloaded deflection of 19 $\mu$m. It required 19 $\mu$N of force to force the actuator back to a zero deflection.

As an alternative to the standard electrical connection, where current flows through the narrow and wide arms in series, the narrow arm and wide arm can be connected in parallel [7]. In this case, the roles of the arms are reversed, with the wide arm being the hot arm. This occurs since the resistance of the hot arm is lower, and so draws more current then the narrow arm. This is obviously not useful for single actuators, since an additional electrical connection would then be required. However, it is useful in more complicated ETC devices, where the actuators are only part of the function. Moulton and Ananthasuresh [7] label this embedded electro-thermal-compliant actuation.

### 12.2.1 Vertical Actuation

Guckel actuators that move perpendicular to the substrate surface are also possible [1] if the surface-micromachining technology has multiple structural layers. The hot and cold arms are created in separate layers. Placing the hot-arm above the cold-arm will lead to actuators that deflect downwards, while placing the hot-arm below the cold-arm will lead to actuators that deflect upwards.

### 12.2.2 Back-bending

Initially, as the electrical power is increased the deflection also increases. However, these actuators obviously reach a limit where further deflection does not occur with increasing electrical power.

If Guckel actuators are driven with electrical power beyond their maximum deflection, the actuators undergo plastic deformation [1]. At increased tem-

*Figure 12.3.* SEM of a team of Guckel actuators. Fabricated using PolyMUMPs™.

peratures, the material becomes plastic[6], and so under stress undergoes irreversible deformation. The hot-arm thus irreversibly buckles under the force of pushing against the cold-arm. This deformation occurs primarily in the hot-arm, which bows [1].

If the input electrical power is increased further, the hot-arm undergoes further plastic deformation. Eventually, the hot-arm thins too severely and burns out [1], leading to device destruction.

Even if back-bending does not occur immediately as described above, long term creep can lead to back-bending over time. Thus, long-term operation of Guckel actuators requires maximum operating powers below that necessary for immediate back-bending. With a proper safety margin, several million cycles are easily realizable [1].

The plastic deformation, or creep, effectively shortens the hot-arm [1]. When power is removed, the actuator does not return to its original position, but has a 'negative' deflection. Thus, back-bending, while a reliability issue, can also be beneficial to create one-time deflections that do not require steady-state power to be maintained. However, if the actuator must still be actuated, it will work, although starting from a 'negative' deflection.

Note that if an actuator is pushing against a stiff structure, the hot-arm may buckle rather then the actuator deflecting. However, without electrical overstressing, this is a reversible deformation and not back-bending.

Guckel actuators can provide more force when back-bending [1]. For example, the actuator fabricated by Comtois and Bright [1] could provide 15.5 $\mu$N in back-bending.

### 12.2.3 Actuator Teams

When the force produced by a single Guckel actuator is insufficient, then several actuators can work in concert. Longer actuators could be used, but that increases the actuator's resistance as well as chance for failure due to stiction

(a) Bent-beam actuator with a single beam.

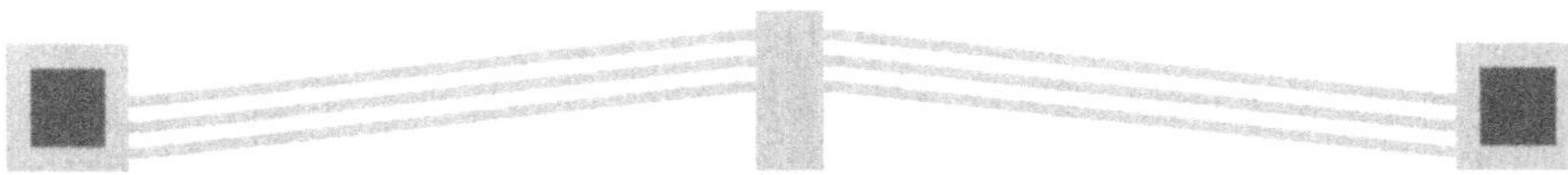

(b) Bent-beam actuator with a three beams.

*Figure 12.4.* Illustration of bent-beam electro-thermal actuators.

[1]. Also, longer actuators do not necessarily produce more force since they are more susceptible to bowing [1].

In micromachining processes containing a single layer, the actuators can be connected together using compliant structures (section 10.1). With a sufficient number of structural layers in the surface-micromachining process, more complex yokes can be designed, although compliant yokes are both simple and compact [1].

Often, Guckel actuators are connected together in symmetric pairs. The actuators are connected using flexible segments to a central shuttle. An important result of connecting two opposing rows of actuators into a team is that the transverse motion created by the arcs cancel [1]. Because the forces due to the unwanted deflection cancel, the central shuttle travel along a true line. and the deflection of the combined motion is linear.

Further, with proper design, each actuator has its own compliant connection to the team. Each additional actuator therefore provides a fixed incremental force [1]. Achieving a desired force consists of simply adding enough actuators to the team.

Electrically, the actuators are connected in parallel[7]. This configuration is stable as long the actuators' resistances increase with temperature. This provides negative feedback to the current flowing through the actuators and so stabilizes the distribution of current throughout the team.

## 12.3 Bent-beam Actuators

Bent-beam actuators, also known as Chevron actuator, rely on the buckling of a straight beam. To control the direction of actuation, the beam is kinked to provide a preferred direction to the buckling. When electric current is passed along the bent beams, thermal expansion pushes the apex outward [2].

One of the key advantages of bent-beam actuators over Guckel actuators is that the motion of the apex is straight [2]. Although it can be corrected, the tip of Guckel actuators actually follow arcs (section 12.2).

To cause actuation, the temperature difference in bent-beam actuators is between the actuator and the substrate. Bent-beam actuators do not have to be differentially heated, and so do not need to be heated ohmically. Thus, they can be actuated using other sources of heat [1].

If unloaded, the maximum displacement of the actuator can be determined using the following equation [2]:

$$d_{\max} = 2\frac{\tan\theta}{k}\tan\left(\frac{kL}{4}\right)\frac{L}{2}\tan\theta \qquad (12.8a)$$

with,

$$k = \sqrt{\frac{F}{EI}} \qquad (12.8b)$$

$$F = \frac{(\Delta T\alpha L - \Delta L')AE}{L} \qquad (12.8c)$$

$$L' = \frac{\tan^2\theta}{4k}(2G + kL + kLG^2 + \sin kL$$
$$-2G\cos kL - G^2\sin kL) \qquad (12.8d)$$

$$G = \tan\left(\frac{kL}{4}\right) \qquad (12.8e)$$

$$H = \tanh\left(\frac{kL}{4}\right) \qquad (12.8f)$$

In equation 12.8a, $d_{\max}$ is the maximum displacement, $\theta$ is the bending angle of the beams[8], $L$ is the length of the actuator as measured straight between the beam ends, $L'$ is the real beam length as measured along the beam path, $E$ is Young's modulus, $\alpha$ is the coefficient of thermal expansion, $A$ is the beam cross-sectional area, $F$ is the reaction force along the x-axis at the beam ends, $\Delta T$ is the temperature change, and $\Delta L'$ is the change in $L'$ due to $F$.

Equation 12.8a suffers from the approximation that the coefficient of thermal expansion is temperature independent [2]. Unfortunately, ETC actuators typically experience a wide temperature range and this assumption does not hold true. For example, the coefficient of thermal expansion varies for silicon varies from $2.5 \times 10^{-6}$ /°K at room temperature to $4 \times 10^{-6}$ /°K at 500°K [2]. This leads to underestimating the displacement by approximately 30%.

Equation 12.8a is accurate to within 5% of finite element analysis [2]. However, its complexity does not encourage casual use, and in most beneficial is relating design parameters to device performance.

*Figure 12.5.* SEM of two bent-beam actuator teams. The top actuator is connected in a team of two and is pulling on a serpentine spring. The bottom actuator is also connected in a team of two, but not connected to anything. Fabricated using PolyMUMPs™.

The maximum loading force, defined as the force necessary to force the actuator's displacement back to zero, can be derived using the complementary energy method:

$$f_{\max} = d_{\max} k_y \tag{12.9}$$

The actuator's stiffness can be calculated as follows:

$$k_y = \frac{4AE \sin^2 \theta'}{L} \tag{12.10}$$

Above, the variable $\theta'$ was introduced. It is the effective bending angle, and is not necessarily equal to $\theta$ when the actuator tip is displaced [2]. In particular, long beams with shallow bending angles will be sensitive to differences in $\theta$ and $\theta'$.

Typical designs values for bent-beam actuators are bending angles of 0.025 to 0.2 radians [2]. Lengths are typically several hundred microns.

Like Guckel actuators, maximum operating voltages are usually less than 10 volts [2], with lower maximum voltages being common. This makes bent-beam actuators compatible with standard CMOS voltages. Further, the thermal isolation of the actuator is on the order of 1000 to 10000 °K/W. Temperature of several hundred degrees require power in the tens of milliwatts [2].

### 12.3.1 Actuator Teams

Bent-beam actuators are particularly easy to connect into actuator teams. The beams can run parallel to each other, and so occupy very little space (figure 12.5). Further, placing the beams together reduces heat loss, and so the efficiency of the individual beams is improved.

A common use of bent-beam actuator is to use two teams of bent-beam actuators to push on each end of an open toggle (section 10.5.2). Bent-beam actuators have relatively small displacements, and so often require mechanical amplification before their output can be used.

## 12.4 Topological Optimization

In addition to the fixed ETC actuator designs just discussed, ETC actuators can also be designed numerically. The design problem is formulated as a minimization problem, and the actuator design is provided by a computer.

Assuming that geometric changes do not influence convection or conduction properties, the electrical and thermal response of the ETC actuator can be modelled as a linear system [8]. In addition to removing strong non-linearities due to contact, this also means that the various physical phenomena are weakly coupled. This allows the electrical, thermal, and mechanical responses to be solved separately[9]. Still, the mathematical description involves 12 equations grouped into three physical domains [8].

Although in surface-micromachining material density is fixed, the optimization method allows the material density to vary. Typically, the analysis involves a relative density, $\rho_r \in [0, 1]$ which varies, and the true material density, $\rho$, which is fixed [8]. This is necessary as many optimization techniques require design variables that are continuous [8].

For the simulations, the material properties are not simply linearly interpolated between zero and the material's true properties. Instead, the properties vary by the relative density raised to some power. The exception is Poisson's ratio, which is independent of the relative density:

$$\sigma_e = \rho_r^p \sigma_e 0 \quad (12.11a)$$

$$\sigma_t = \rho_r^p \sigma_t 0 \quad (12.11b)$$

$$E(\rho_r) = \rho_r^p E_0 \quad (12.11c)$$

$$\nu(\rho_r) = \nu_0 \quad (12.11d)$$

Above, the electrical conductivity, $\sigma_e$, the thermal conductivity, $\sigma_t$, Young's Modulus, $E$, and Poisson's ratio, $\nu$, depend on the relative densitiy. For all material properties, the interpolated properties equal the true material property when $\rho_r = 1$.

There are constraints on the power-law exponent, $p$. First, the power-law exponent must satisfy $p \geq 2$ for the electrical and thermal conductivity. Further, for the mechanical simulation to be thermodynamically admissible, the following constraint must be met:

$$p \geq \max\left(\frac{2}{1-\nu_0}, \frac{4}{1+\nu_0}\right) \tag{12.12}$$

The area that the actuator will occupy is broken up into small regions [8]. The relative density is allowed to vary in each region. The electrical, thermal, and mechanical responses of the actuator can then be modelled. For example, the discretized model can be used in finite element analysis [8]. Evaluation of the system simulation is used as a cost function for the optimization.

With the system model in hand, one must determine the optimization problem. The details of the formulation and solution can be found in Sigmund [8].

$$\begin{aligned}
\max_{\rho} \quad & u_{\text{out}}(\rho) \\
\text{subject to:} \quad & \\
& \sum_{e=1}^{N} \rho_{r,e} V_e \leq V^*, \\
& I(\rho) \leq I^*, \\
& \frac{\hat{u}_{\text{out}}^2(\rho)}{u_{\text{out}}(\rho)} \leq \epsilon^*, \\
= \quad & \mathbf{0} < \rho_{\min} \leq \rho \leq \mathbf{1}
\end{aligned}$$

Above, $V_e$ is the element volume, $V^*$ is a constraint on the total material volume, $I(\rho)$ is the electrical current, $I^*$ is the maximum current, $\rho_{\min}$ is the minimum relative density, $\hat{u}_{\text{out}}$ is the actuator motion at the output perpendicular to the desired direction, $u_{\text{out}}$ is the actuation motion at the output, and $\epsilon^*$ is a small positive number.

This method can be extended to multiple inputs and outputs [8]. For example, one could choose to maximize $u_{\text{out},1} + u_{\text{out},2}$ for a system with two outputs. However, additional constraints are also necessary.

## Notes

1 This statement is really 'larger force per unit area.' Force can be increased by using actuators in teams.

2 These actuators are also known as pseudo-bimorph actuators, or standard thermal actuators.

3 The shape factor corrects for shape dependence on the heat flow.

4 This does not mean never.

5 The coefficient of thermal expansion (CTE) is not temperature independent, although this approximation is often made. For example, the CTE of silicon increases by 65% between room temperature and 700°C.

6 For example, silicon is a brittle material at room temperature. However, at higher temperatures it becomes plastic.
7 Other type of electrical connection are possible, and discussed by Comtois and Bright [1].
8 If the small angle at the apex is $\alpha$, then $\theta = \pi/2 - \alpha$.
9 Clearly, information from the simulations is used to update following simulations. This is also known as sequential simulation.

## References


[1] J.H. Comtois and V.M. Bright. Applications for surface-micromachined polysilicon thermal actuators and arrays. *Sensors and Actuators A*, 58(1):19–25, 1997.

[2] L. Que. Bent-beam electrothermal actuators - part i: Single beam and cascaded devices. *jmems*, 10(2):247–254, 2001.

[3] H. Guckel, D. Burns, C. Rutigliano, E. Lovell, and B. Choi. Diagnostic microstructures for the measurement of intrinsic strain in thin films. *Journal of Micromechanics and Microengineering*, 2(2):86–95, 1992.

[4] J.H. Comtois, M.A. Michalicek, and C.C Barron. Electrothermal actuators fabricated in four-level planarized surface micromachined polycrystalline silicon. *Sensors and Actuators A*, 70(1-2):23–31, 1998.

[5] Q.-A. Huang and N.K.S. Lee. Analysis and design of polysilicon thermal flexure actuator. *Journal of Micromechanics and Microengineering*, 9(1):64–70, 1999.

[6] Q.-A. Huang and N.K.S. Lee. Analytical modeling and optimization for a laterally-driven polysilicon thermal actuator. *Microsystem Technologies*, 5(3):133–137, 1999.

[7] T. Moulton and G.K. Ananthasuresh. Micromechanical devices with embedded electro-thermal-compliant actuation. *Sensors and Actuators A*, 90(1-2):38–48, 2001.

[8] O. Sigmund. Design of multiphysics actuators using topology optimization - part i: One-material structures. *Computer Methods in APplied Mechanics and Engineering*, 190(49-50):6577–6604, 2001.

# Chapter 13

# RAISED STRUCTURES

Surface-micromachined devices are typically contained in a volume specified by the area of the chip and the height of the thin-films used during fabrication. Even a very small chip will be several millimetres on a side. However, the total height of the thin-films will typically be measured in microns. Thus, surface-micromachined devices are quite constrained in this one direction[1]. This small vertical range can be a disadvantage.

However, even a minimal surface-micromachining process allows the construction of hinges. This opens the possibility of moving components out of the very short volume otherwise allowed. Hinges allow true three-dimensional structures to be fabricated using surface-micromachining [1–3], which is otherwise a planar process.

In raised structures, also known as flip-up structures, parts are typically fabricated flat and connected to the substrate with hinges. After the release step, the parts can then be rotated into position. Since raised structures are free to move after release, it may be necessary to use tethers (section 8.4) to constrain their motion during the release process.

Application of raised structures include sensors [1], radio frequency devices [4], and, particularly, optical devices [1, 3–7].

## 13.1 Hinges

### 13.1.1 Torsion Hinges

Torsion hinges use a long beam to allow rotation. The rotation is applied as a torque, twisting the beam (section 5.2). This technique is important in a

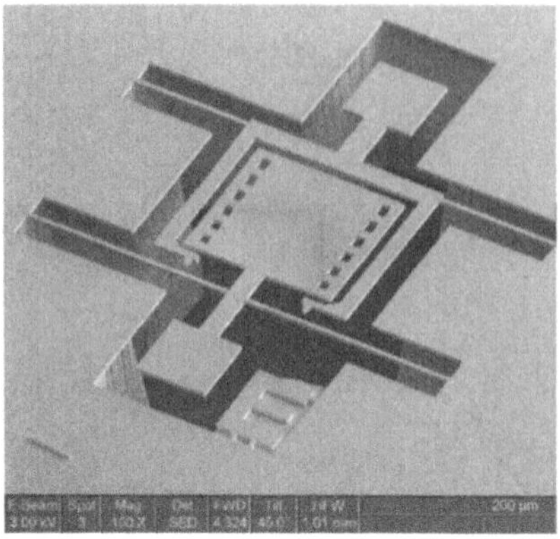

*Figure 13.1.* SEM of a prop-up mirror before assembly. Fabricated using SoiMUMPs™.

*Figure 13.2.* SEM of a torsion hinge used in a prop-up mirror. Fabricated using SoiMUMPs™.

number of microfabrication technologies because it can be implemented using a single structural layer [8].

In technologies with more than one structural layer, torsion hinges are not very common. First, they require significantly more space than the other types of hinges. However, and probably more importantly, the length necessary to reduce the torsional stiffness, also reduces the stiffness to other transverse loads, such as transverse forces and bending.

Torsion hinges are still necessary in technologies with multiple structural layers. One reason they may be preferred is that, unlike the other hinges described below, torsion hinges are physically continuous. This is important when the raised structure requires electrical access [1].

The length necessary for torsion hinges can be reduced by folding the torsion hinge. For example, a serpentine springs or an 'H'-springs may be used, although the designs need to be modified to avoid interference with the substrate.

### 13.1.2 Staple Hinges

If there are two structural layers, then it is possible to replace torsion hinges with staple or scissor hinges. The simplest is the staple hinge [1]. The hinge consists of two parts: the axel, which is constructed from the first structural layer, and the staple, which is constructed from the second structural layer. Typically, the staple is attached to the substrate surface at two points (figure 13.3), completely enclosing the axel and thus trapping it to the wafer surface.

Staple hinges provide a complete 180° of freedom. In fact, device motion is not limited by the hinge, but by the presences of the substrate.

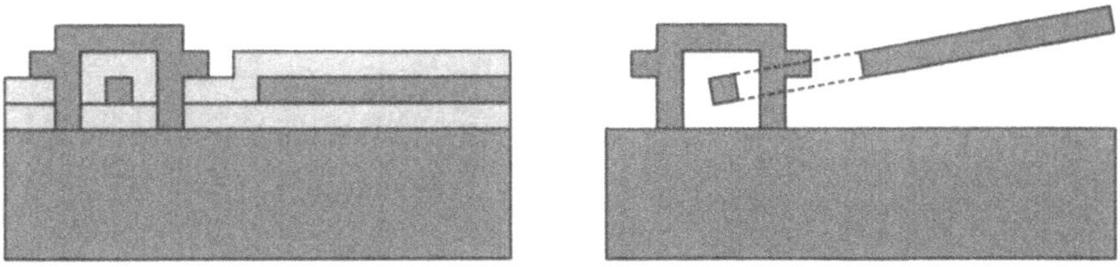

(a) before release (b) after release

*Figure 13.3.* Cross-section of a simple staple hinge design, both before and after release. Medium grey represents the substrate and the structural material, while the sacrificial material is represented by light grey.

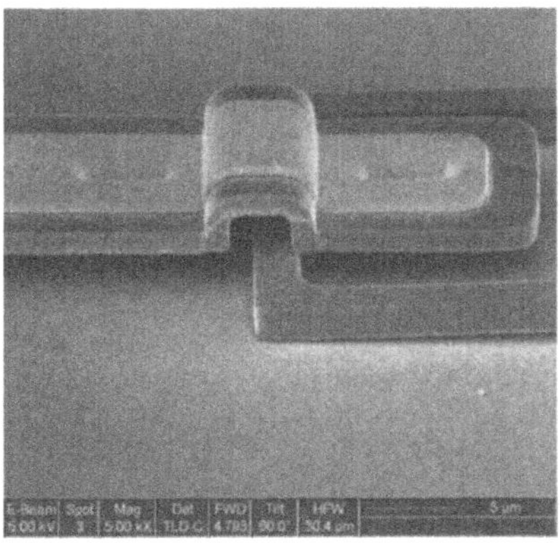

*Figure 13.4.* SEM of a simple staple hinge. Fabricated using PolyMUMPs™.

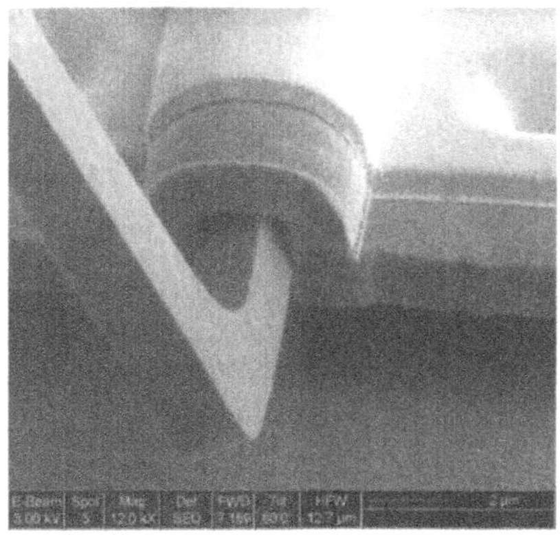

*Figure 13.5.* SEM of a staple that has jammed. Fabricated using PolyMUMPs™.

While very useful, it is important to note that staple hinges are constructed from essentially rectangular components. This makes the hinges less than ideal. It is common for them to become trapped (figure 13.5). Careful design and refinement are often required to increase their reliability.

A variant of the staple hinge is to anchor the staple at only one location [9]. With conformal deposition, the topography (section 7.1.4) of the second structural layer will still constrain the motion of the axel. Friedberger and Muller [9] also investigated the use of compliant structures to guide and constrain motion of the hinge during deployment. In particular, they designed the hinges so that components were help upright once raised.

## 13.1.3 Scissor Hinges

It is not necessary to use the substrate when creating hinges. Scissor hinges use only two structural layers, and can create a hinge between two plates [1]. In scissor hinges, connecting bars are used to thread an interdigitated structure.

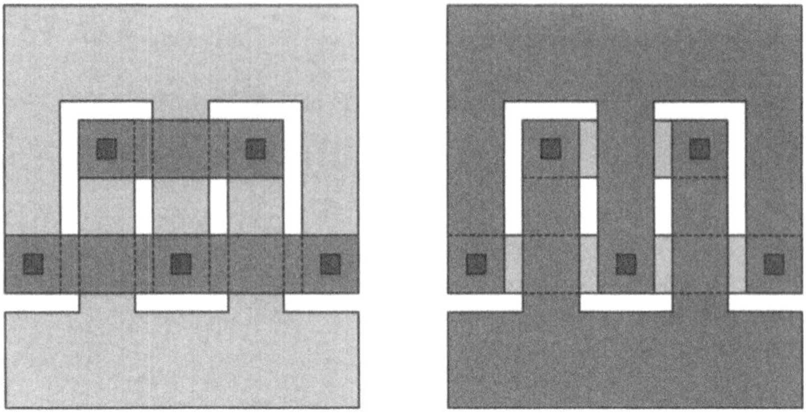

(a) Downward deflecting hinge (b) Upward deflecting hinge

*Figure 13.6.* Illustration of the layouts for both types of scissor hinges. Light grey is the bottom structural layer, medium grey is the top structural layer, and dark grey is a via connecting the structural layers.

*Figure 13.7.* SEM of a scissor hinge. Note that the hinge can only fold upwards. Fabricated using PolyMUMPs™.

If the threads pass over the fingers, then the hinge will allow structures to rotate closing downwards, while if the threads pass under the fingers, then the hinge will allow structures to rotate closing upwards.

## 13.2 Assembly

Additional manufacturing steps are required for raised structures over and above the surface-micromachining. After the release step, the structures are still lying flat on the wafer surface, and still need to be assembled.

The methods of assembling raised structures discussed below do not form an exhaustive list. Other important methods include the use of polymers [10, 11]

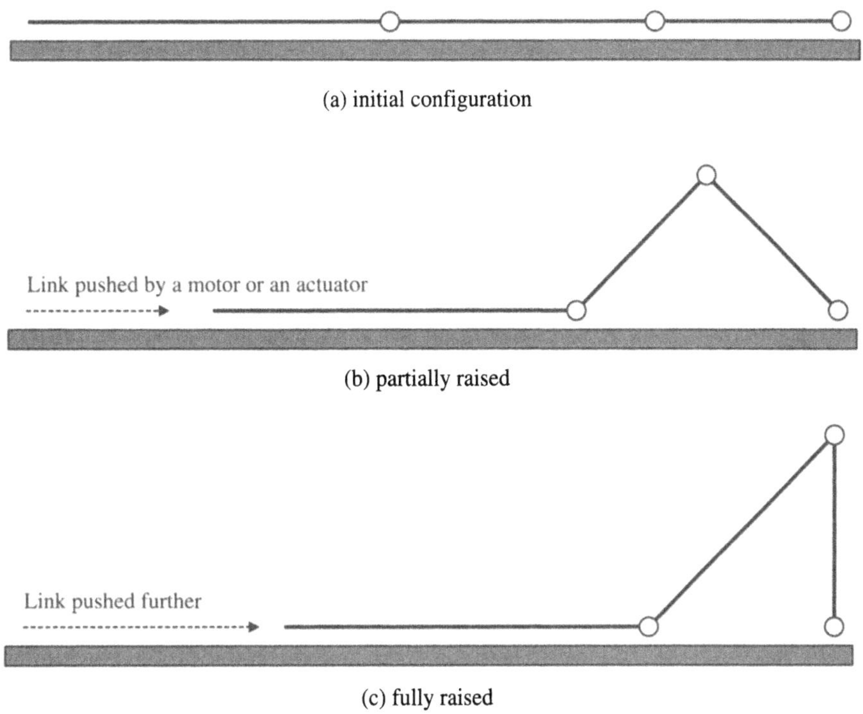


*Figure 13.8.* Illustration of a link configuration that can be used to raise a structure [12].

### 13.2.1 Manual Assembly

The technologically simplest method of assembling raised structures is manual assembly. An operator, using micropositioners, inserts fine probes underneath the components, and then lifts them off the wafer surface. The necessary equipment is present in most microprober stations, which are common pieces of microelectronics testing equipment.

Manual assembly is labour intensive and repetitive [1,12], and so automated solutions are preferred [3, 4, 12]. Automated assembly can reduce the assembly time and cost [3]. Assembly typically requires 10 minutes of labour per structure.

An additional disadvantage of manual assembly is that it is a sequential process. Raised structures are assembled one at a time. For systems integration, where many raised structures may be required, assembly could become a manufacturing bottleneck.

### 13.2.2 Microactuator Assembly

Many types of actuators can be built using surface micromachining (chapters 11 and 12). Instead of requiring off-chip motion to assemble the raised

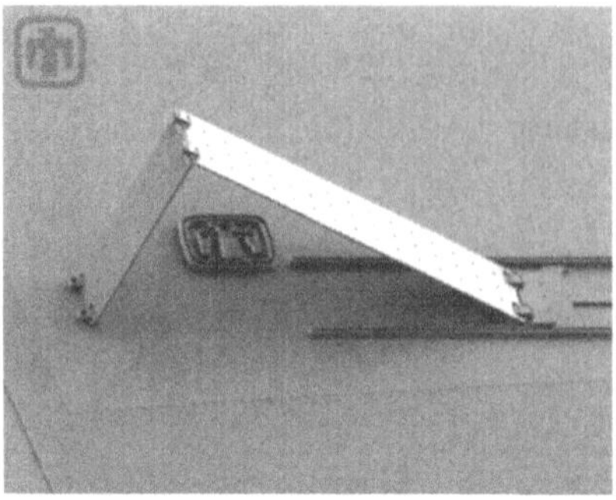

*Figure 13.9.* A mirror system design; in this system the mirror is elevated by a three-gear torque-multiplying system. The mirror is shown in the upright position. Fabricated using SUMMiT™. Courtesy Sandia National Laboratories, SUMMiT™ Technologies, www.mems.sandia.gov.

structures, actuators and motors can be built on-chip to perform the necessary actions [7, 12, 13].

An example configuration is shown in figure 13.8. In consists of three links connected by hinges [7, 12]. There is also a motor or an actuator connected to the left side on the structure. Pushing the leftmost link to the right raises the structure.

Care must be taken to provide an initial bend in the structure so that it will move upwards. This can be accomplished a number of ways. Reid *et al* [12] used a vertical actuator to lift the structure part of the way before engaging the motor.

To obtain the necessary force or torque to complete the assembly process, large actuators may be required. Microactuator assembly thus consumes additional chip area, which is disadvantageous [4]. However, if the resulting devices need to be actuated when in use, then the additional chip area is not a major drawback.

### 13.2.3 Surface Tension Assembly

Surface tension assembly relies on the fact that, for miniature components, surface tension is often stronger than gravity. Thus, surface tension can be used to move miniature components.

The assembly of components using surface tension was first demonstrated using solder [2,3]. The solder is placed at the hinges of components. When the solder melted, it pulls the parts into position. Materials other than solder can be used, and if the reflow material can be patterned using photolithography, surface tension assembly can be a batch process.

For example, borophospho-silicate glass is another material used in surface tension assembly [3]. The viscosity of the glasses can be controlled. Thus, different axis can be programmed to assemble at different rates, allowing more complex assembly sequences to be constructed. However, the borophospho-silicate glasses melt higher temperatures than solder.

To obtain structures that move to and angle of 90° with respect to the substrate, the amount of reflow material must be properly selected. If the reflow material is patterned, then the width of the reflow material, called the driver pad, can be selected based on the following equation [3]:

$$w = \frac{8t}{\pi + 2} \quad (13.1)$$

Above, $w$ is the width of the driver pad on each side of the hinge[2], and $t$ is the thickness of the reflow material.

If angles other than 90° are desired, then the size of the driver pads can be modified. Unfortunately, relying on the volume of the deposited reflow material is unreliable. Instead, the preferred method is to use mechanical stops control the angle of the raised components.

### 13.2.4 Thermokinetic Assembly

Thermokinetic assembly relies on the momentum imparted to a structure when atoms and molecules strike a solid surface. In the molecular flow regime[3], gas molecules do not interact, and so do not thermalize. Instead, most gas molecules travel across the reaction chamber without colliding with anything. The gas molecules thus travel with their velocity determined by the kinetic energy imparted by the last surface they interacted with.

### 13.2.5 Magnetic Assembly

Components can also be raised by magnetic forces. Typically, the components interact with a uniform global magnetic field created off-chip.

If additional materials can be introduced into the fabrication process, a structural layer made from a magnetic material can be introduced. This structural layer can then be patterned into strips along the top edges of components to be raised [14]. When a magnetic field is imposed on the chip, the permalloy interacts with the external magnetic force creating an upwards force. This creates a torque about the hinges, lifting the structures off the substrate surface.

An advantage of this process is that it consumes no extra chip space [14]. The magnetic material is patterned directly atop the existing structures. A second advantage is that multiple devices can be actuated in parallel, which is important in batch fabrication. This is, however, also a disadvantage as it is difficult to control components individually.

Assuming that the magnetic material is rectangular, the torque can be calculated as follows [14]:

$$T_m = M w_m t_m l_m H_{ext} \cos\theta \quad (13.2)$$

In equation 13.2, $T_m$ is the torque acting on the hinged structure, $M$ is the magnetisation of the structure, $w_m$, $t_m$, and $l_m$ are the width, thickness, and length of the structure, $H_{ext}$ is the external magnetic field, and $\theta$ represent the angle of the structure with respect to the substrate. When performing this analysis, designers should pay attention to the saturation magnetization of the material, $M_s$ [14].

Yi and Liu [14] extended their process to raised structure to angles other than 90°. This was performed by using cantilevers as mechanical stops.

Instead of introducing new materials, surface-micromachined components can interact with the external magnetic field through the Lorentz force. The advantage of this method is that no additional materials or layers need to be added to the manufacturing process. However, it can be difficult to get good electrical connections to components which are constrained solely by mechanical hinges.

## Notes

1 Additionally, there is no photolithographic control over features perpendicular to the wafer surface.

2 The total width of the driver pad is therefore $2w$.

3 The molecular regime, for air, occurs at pressures less than 10 Torr [4].

## References


[1] K.S.J. Pister, M.W. Judy, S.R. Burgeett, and R.S. Fearing. Microfabricated hinges. *saa*, 33(3):249–256, 1992.

[2] R.R.A. Syms and E.M. Yeatman. Self-assembly of three-dimensional microstructures using rotation by surface tension forces. *Electronics Letters*, 29(8):662–664, 1993.

[3] R.R.A. Syms. Surface tension powered self-assembly of 3-d micro-optomechanical structures. *Journal of Microelectromechanical Systems*, 8(4):448–455, 1999.

[4] V. Kaajakari and Amit Lal. Thermokinetic actuation for batch assembly of microscale hinged structures. *Journal of Microelectromechanical Systems*, 12(4):425–432, 2003.

[5] N.C. Tien, O. Solgaard, M.-H. Kiang, M. Daneman, K.Y. Lau, and R.S. Muller. Surface-micromachined mirrors for laser-beam positioning. *Sensors and Actuators A*, 52(1-3):76–80, 1996.

[6] J.H. Comtois and V.M. Bright. Applications for surface-micromachined polysilicon thermal actuators and arrays. *Sensors and Actuators A*, 58(1):19–25, 1997.

[7] L.Y. Lin, E.L. Goldstein, and R.W. Tkach. Free-space micromachined optical switches with submillisecond switching time for large-scale optical crossconnects. *IEEE Photonics Technology Letters*, 10(4):525–527, 1998.

[8] S. Kurth, R. Hahn, C. Kaufmann, K. Kehr, J. Mehner, U. Wollmann, W. Dotzel, and T. Gessner. Silicon mirrors and micromirror arrays for spatial laser beam modulation. *Sensors and Actuators A*, 66(1-3):76–82, 1998.

[9] A. Friedberger and R.S. Muller. Improved surface-micromachined hinges for fold-out structures. *Journal of Microelectromechanical Systems*, 7(3):315–319, 1998.

[10] K. Suzuki, I. Shimoyama, and H. Miura. Insect-model based microrobot with elastic hinges. *Journal of Microelectromechanical Systems*, 3(1):4–9, 1994.

[11] E. Smela, O. Inganas, and I. Lundstrom. Controlled folding of micrometer-sized structures. *Science*, 268(5218):1735–1738, 1995.

[12] J.R. Reid, V.M. Bright, and J.T. Butler. Automated assembly of flip-up micromirrors. *Sensors and Actuators A*, 66(1-3):292–298, 1998.

[13] R. Yeh, E.J.J. Kruglick, and K.S.J. Pister. Surface-micromachined components for articulated microrobots. *Journal of Microelectromechanical Systems*, 5(1):10–17, 1996.

[14] Y.W. Yi and C. Liu. Magnetic actuation of hinged microstructures. *Journal of Microelectromechanical Systems*, 8(1):10–17, 1999.

PART IV

# PROCESSES

Chapter 14

# POLYMUMPS$^{TM}$ PROCESS

PolyMUMPs$^{TM}$ is one of the oldest commercially available surface-micromachining processes. The name is an acronym for polysilicon multi-user micromachining process. It contains one ground plane layer (section 8.1), plus two other structural layers (section 3.3).

The main layers used in the PolyMUMPs$^{TM}$ process, along with the mask names used to control that layer's geometry, are outlined in table 14.1. There are additional masks in the process. For example, a mask is provided in this process, ANCHOR2, that is used instead of combining P1P2VIA and ANCHOR1 when connecting POLY2 to the substrate.

PolyMUMPs$^{TM}$ originally just called MUMPs, started at the Berkeley Sensors and Actuators Center, at the University of California. The original work was done in the late 1980's and early 1990's.

The micromachining process begins with (100) silicon wafers. In particular, the starting wafers used in PolyMUMPs$^{TM}$ have resistivities between 1 and

| Material | Thickness | Purpose | Mask |
|---|---|---|---|
| Silicon Nitride | 0.6 $\mu$m | Isolation Layer | |
| Polysilicon | 0.5 $\mu$m | Ground Plane | POLY0 |
| Silicon Dioxide | 2.0 $\mu$m | Sacrificial Layer | ANCHOR1 |
| Polysilicon | 2.0 $\mu$m | Structural Layer | POLY1 |
| Silicon Dioxide | 0.75 $\mu$m | Sacrificial Layer | P1P2VIA |
| Polysilicon | 1.5 $\mu$m | Structural Layer | POLY2 |
| Gold | 0.5 $\mu$m | Conduction | METAL |

*Table 14.1.* The layers in the PolyMUMPs$^{TM}$ process.

2 Ωcm. The surface of the wafer is heavily doped with phosphorus, further reducing the resistivity. The isolation layer is then deposited, and is formed from 600 nm of low-stress silicon nitride. At this point, the first structural layer is deposited and a surface-micromachining process begins (section 3.3). The structural material is polycrystalline silicon, and the sacrificial material is silicon dioxide.

Additionally, the process includes a gold layer after POLY2. This layer is not strong enough to be used as a structural layer, and is present solely to create conducting lines with high conductivities.

## 14.1 Design Rules

The complete design rules can be found in the Design Handbook [1].

In general, the minimum feature size is 2 $\mu$m. However, the nominal feature size is 3 $\mu$m. The largest constrained dimension is 5 $\mu$m, and controls the enclosure of POLY2 over ANCHOR2.

## 14.2 Photo Gallery

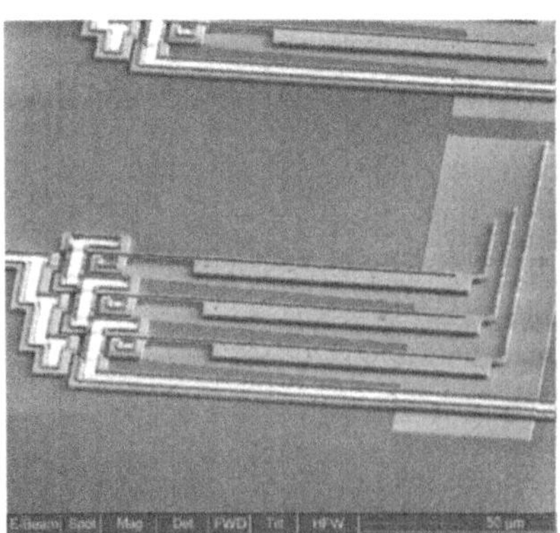

Three Guckel actuators. They are 200 $\mu$m long.

A device that uses a team of bent-beam actuators.

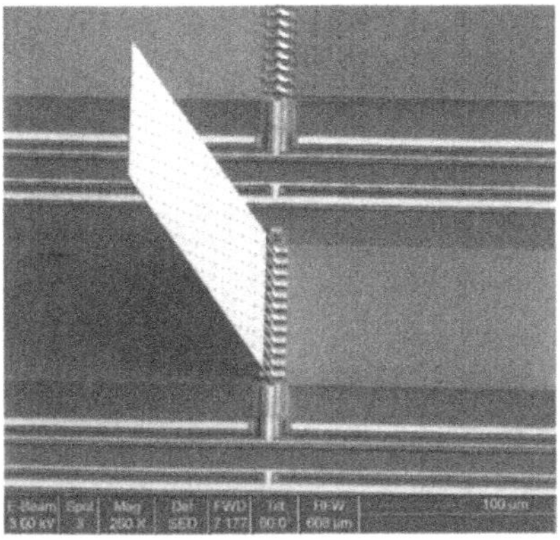

A raised 300 $\mu$m by 300 $\mu$m plate.

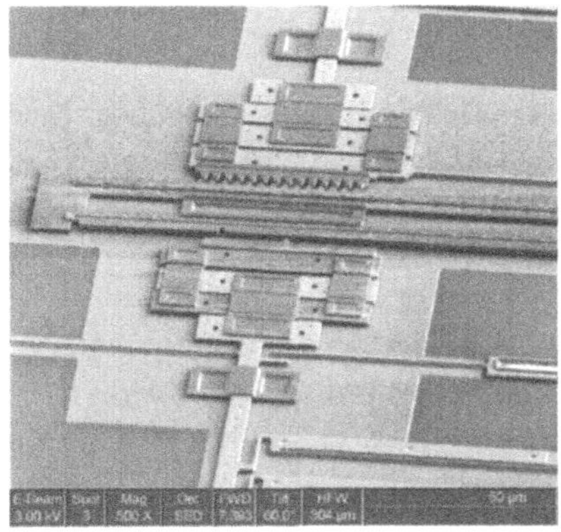

A stepper motor.

# References

[1] D. Koester, A. Cowen, R. Mahadevan, and B. Hardy. *PolyMUMPs Design Handbook.* MEMSCAP, http://www.memsrus.com/mumps.pdf, 2002.

# Chapter 15

# SUMMIT$^{TM}$ PROCESS

This chapter covers both the original SUMMiT$^{TM}$ process and the newer SUMMiT V$^{TM}$ process. Both processes are surface micromachining processes using polycrystalline silicon as the structural material. The main difference between the two processes is that the SUMMiT V$^{TM}$ process contains five polysilicon layers, compared to four in the original process. In both processes, the first layer serves as a ground plane.

The name SUMMiT is an acronym for Sandia Ultra-planar, Multi-level MEMS Technology.

This process is similar to PolyMUMPs$^{TM}$, except that it contains two improvements. First, the SUMMiT$^{TM}$ processes contain one or two additional structural layers, which make possible several structures not possible in a three layer surface-micromachining process. Another important difference is that the

| Material | Thickness | Purpose |
|---|---|---|
| Silicon Dioxide | 0.63 $\mu$m | Isolation Layer |
| Silicon Nitride | 0.8 $\mu$m | Isolation Layer |
| Polysilicon | 0.3 $\mu$m | Ground Plane |
| Silicon Dioxide | 2.0 $\mu$m | Sacrificial Layer |
| Polysilicon | 1.0 $\mu$m | Structural Layer |
| Silicon Dioxide | 0.5 $\mu$m | Sacrificial Layer |
| Polysilicon | 1.5 $\mu$m | Structural Layer |
| Silicon Dioxide | 2 $\mu$m | Sacrifical Layer |
| Polysilicon | 2.25 $\mu$m | Structural Layer |

*Table 15.1.* The layers in the SUMMiT$^{TM}$ process.

sacrificial layers before the fourth and, if available, fifth structural layers are planarized. This ensures that that following structural layers are flat, avoiding interference faults (section 7.1).

## 15.1 Design Rules

The minimum feature size in SUMMiT$^{TM}$ is 1 micron.

## 15.2 Photo Gallery

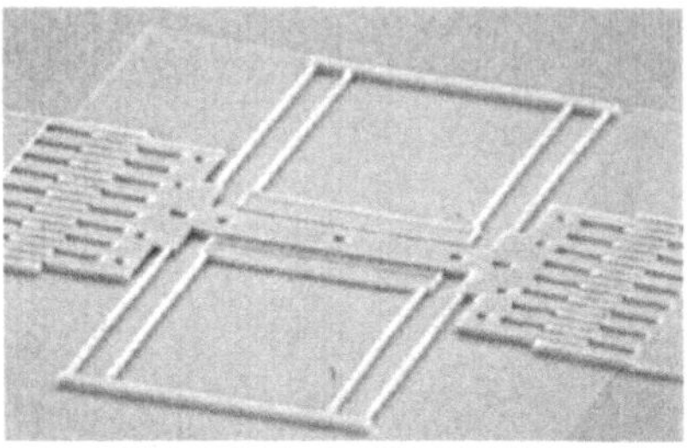

A comb drive actuator fabricated using the ground plane and the first structural layer. Courtesy Sandia National Laboratories, SUMMiT$^{TM}$ Technologies, www.mems.sandia.gov.

Several intermeshed gears fabricated from the first structural layer. Courtesy Sandia National Laboratories, SUMMiT$^{TM}$ Technologies, www.mems.sandia.gov.

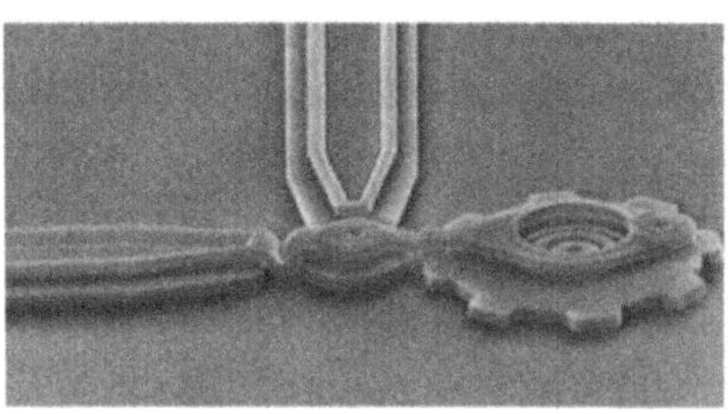

The third structural layer makes it possible to link an actuator to a tool. Courtesy Sandia National Laboratories, SUMMiT$^{TM}$ Technologies, www.mems.sandia.gov.

The torque delivered by the engine is amplified by the transmission. The engine and transmission drive a linear rack that elevates a hinged mirror. Courtesy Sandia National Laboratories, SUMMiT$^{TM}$ Technologies, www.mems.sandia.gov.

# Chapter 16

# MUSIC® PROCESS

Unlike the other two surface-micromachining processes discussed, the structural material in MUSiC® is a silicon carbide.

Another interesting difference in the MUSiC® process is the use of micromoulding layers. In this process, a sacrificial mould material is deposited, and the 'reverse' geometry is etched into the mould material. The structural material, silicon carbide, is then deposited, and the wafer is polished back using chemical-mechanical-polishing (section 2.6) until the top of the mould is exposed. Thus, the structural material exists only in the mould, and following depositions are always atop a flat surface.

Because this process uses micromoulding, there is no difference between sacrificial layers and structural layers. In principle, the final wafer should present a number of disjoint vertical regions. However, the mask conventions were selected to mimic PolyMUMPs™, and so hide this fact.

| Material | Thickness | Purpose | Mask |
|---|---|---|---|
| Silicon Nitride | 0.6 $\mu$m | Isolation Layer | |
| Silicon Carbide | 0.5 $\mu$m | Ground Plane | SIC0 |
| Silicon Carbide (Silicon Oxide Mold) | 2.0 $\mu$m | Structural Layer | ANCHOR1 (reverse field) |
| Silicon Carbide (Polysilicon Mold) | 2.0 $\mu$m | Structural Layer | SIC2 |
| Silicon Oxide | 0.75 $\mu$m | Sacrificial Layer | SIC2_SIC3_VIA |
| Silicon Carbide | 1.5 $\mu$m | Structural Layer | SIC3 |
| Nickel | 0.75 $\mu$m | Conduction | METAL1 |

*Table 16.1.* The layers in the MUSiC® process.

The process is also unique because a special etching sequence following the SIC2 deposition to create hubs for revolute joints.

Fabrication begins with a p-type wafer, with a resistivity between 1 and 10 Ωcm. The isolation layer is 0.6 $\mu$m of silicon nitride. The surface-micromachining process then begins with the ground plane or shield layer, referred to as SIC0[1].

## 16.1 Design Rules

The minimum feature size and spacing for the structural masks vary between 2.0 $\mu$m and 5.0 $\mu$m, with the nominal values varying between 3.0 $\mu$m and 5.0 $\mu$m [1].

The other masks (vias and dimples) have similar size. However, the ANCHOR2 mask may require 6.0 $\mu$m when used to create revolute joints.

## 16.2 Photo Gallery

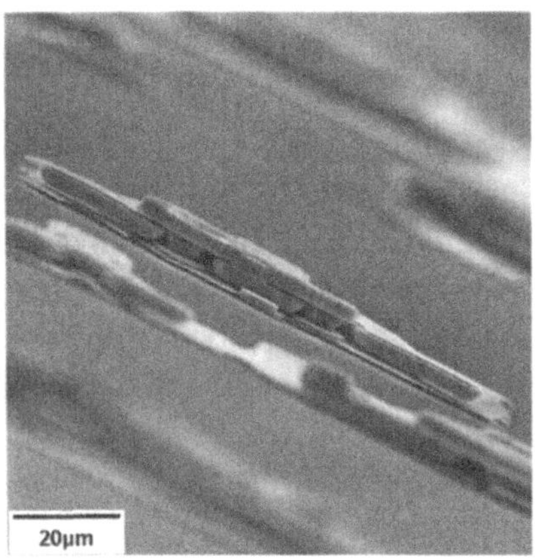


SEM of a cross-section of a wheel.

SEM of four Guckel actuators.

## Notes

1 Note that SIC0 is similar to POLY0 in PolyMUMPs™.

## References

[1] FLX Micro. *Multi-User Silicon Carbide (MUSiC®) Microfabrication Process: Design Handbook.* FLX Micro, Solon, Ohio, 2003.

# GLOSSARY

**Actuator:** A device that creates a motion in response to an input signal.

**Anchor:** A cut or via patterned into a sacrificial layer. Anchors are used to connect structural layers together.

**Bulk micromachining:** A process of creating pits, trenches, cantilevers, and membranes in silicon where large portions of the substrate are removed by etching.

**Chemical Vapour Deposition (CVD):** The formation of a solid film on a substrate by the reaction of vapour phase chemicals which are decomposed or reacted on or near the substrate.

**Deposition:** Refers to the many different processes used to add a thin-film to one or both sides of substrates.

**Epitaxy:** Growing a crystal of one material on the crystal base of another material so that the crystalline orientation of the new material is the same as that of the substrate.

**Lithography:** Process of transferring a pattern. In particular, photolithography transfers a pattern from a mask to a photoresist using a photographic tool (mask aligner), and to the silicon substrate using etching techniques.

**MEMS:** Acronym for micro-electro-mechanical systems. However, the term has come to include all types of microfabricated devices.

**Micromachining:** Process of manufacturing miniature mechanical systems.

**Microfabrication:** All of the technologies and activities associated with manufacturing miniature devices and systems.

**MOEMS:** Acronym for micro-optical-electro-mechanical systems.

**MST:** Acronym for microsystems technology.

**Photoresist:** Material that is used in photolithography, a key process in transferring 2D patterns from a mask to a wafer surface.

**Physical vapour deposition (PVD):** process in which the material to be deposited is converted from a solid phase into vapour phase, then moved through a region of low pressure, with the vapour condensing on the substrate, to form a solid thin film.

**PolyMUMPs™:** Acronym for the polysilicon multi-user micromachining process. It is a three layer polysilicon surface-micromachining process.

**Rotor:** This is the mobile or moving electrode in an electrostatic actuator.

**Sacrificial material:** Material used to create spaces or voids in a surface-micromachining process.

**SoiMUMPs™:** Acronym for the SOI multi-user micromachining process. A hybrid process bulk/surface-micromachining process.

**Stator:** This is the stationary electrode in an electrostatic actuator.

**Structural material:** Material used to create structures that will remain in the final device in a surface-micromachining process.

**Substrate:** Foundation for planar microfabrication processes, particularly surface micromachining and microelectronics. Substrates have one, maybe two, polished sides upon which fabrication takes place.

**SUMMiT™:** Acronym for the Sandia Ultra-planar, Multi-level MEMS Technology. It is a four or five level polysilicon surface-micromachining process.

**Surface micromachining:** Micromachining process that uses the deposition and patterning of multiple thin-films.

**Wafer:** Substrate and all materials deposited during fabrication.

# About the Authors

R.W. Johnstone received a B. Eng & Soc. degree in engineering physics (with honours) from McMaster University, Hamilton, Ontario, Canada, in 1999, and an M. A. Sc. degree in engineering from Simon Fraser University, Burnaby, British Columbia, Canada, in 2002.

M. Parameswaran (M'92) received the B.S. degree in electronics engineering and communication (with honours) from the Madras University, Madras, India, in 1982, and the M.S. degree in electrical engineering and Ph.D. degree in microelectronics from the University of Alberta, Alberta, Canada, in 1990. He then joined the School of Engineering Science, Simon Fraser University, Burnaby, B.C., Canada, as an Assistant Professor, where he established a microfabrication and micromachining laboratory to fabricate micromechanical devices and systems. During the 1998 academic year, he spent his sabbatical leave at the Berkeley Sensor and Actuator Center, University of California at Berkeley, where he performed research in microphotonics and display applications. In June 2000, he was promoted to Full Professor and has also been appointed the Director of the Institute of Micromachine and Microfabrication Research, School of Engineering Science, Simon Fraser University. He has served as a referee and consultant for agencies such as the National Institute of Standards and Technology, the National Institute of Health, NATO, and the United Nations. He has authored or co-authored over 80 technical documents in scientific journals and conferences.

# Index

**Batch number: 10584448**

Printed by Printforce, the Netherlands